ALTERNATIVE MOTOR FUELS

A NONTECHNICAL GUIDE

ALTERNATIVE MOTOR FUELS

A NONTECHNICAL GUIDE

MAUREEN SHIELDS
LORENZETTI

NEW ENGLAND INSTITUTE
OF TECHNOLOGY
LEARNING RESOURCES CENTER

PENNWELL NONTECHNICAL SERIES

#34742728

11/00

ACKNOWLEDGEMENTS

This book was only possible through the help and advice of many individuals and groups dedicated to safe and reliable transportation from alternative sources. A special thanks is in order to the U.S. Alternative Fuels Hotline and the Energy Information Administration, which graciously allowed me to reprint some charts on various alternative fuel properties. I also would like to thank my long-suffering editor, Marla Patterson, for her valiant efforts to meet publishing deadlines, and to my husband, Stephen, and daughter, Gina, for their patience.

In Memory of Dr. Brian Davis, Sun Company

Educator, Colleague, and Friend

C O N T E N T S

F O R E W O R D

Those who have followed the evolution of U.S. motor fuels from the introduction and later phaseout of lead, to reformulated gasoline and alternative fuels know the business of making fuel is more complicated than ever. Growing concerns by the public and regulators over the quality of our nation's air and the security of energy supplies has led to a series of laws that will mean staggering growth for nonpetroleum fuel sources this decade and beyond.

But what exactly are alternative fuels? Alcohols, natural gas, hydrogen, and even solar are examples, and each has its own advantages and disadvantages, depending on issues such as price, region, and availability of supply. As a result, this book was written to provide two basic, necessary services to readers: a reference guide for industry veterans and a teaching tool to those new to the industry.

Despite recent government efforts to trim back environmental laws, alternative fuels are not a passing fancy of environmentalists or regulators. They are required by law and will be phased in to replace a small, but significant percentage (5-10 percent) of petroleum motor fuels this century.

If a fuel supplier, fleet manager, or government agency opts to use alternative fuels, whether for legal or economic reasons, the marketplace offers confusing signals. To help sort out which fuel may fit a user's specific needs, each fuel chapter outlines the basic factors needed to make an informed decision, including supply and pricing issues, safety considerations, storage and handling, and air quality benefits. There also is a chapter devoted to fleet considerations surrounding alternative fuel vehicles (AFV), such as resale value, range, and cost compared to conventional vehicles. Finally, the book includes appendices providing details on various laws impacting the AFV market and other clean fuel compliance issues.

CHAPTER ONE

INTRODUCTION

The Pace of Progress: Society is a mule, not a car — if pressed too hard, it will throw off its rider.

— Anonymous

If some clean air idealists had their way, motorists would never spend another cent on gasoline. Gas stations would disappear. In their place would be "recharging" centers at the local shopping mall, home, or office where a vehicle could be refueled with electric power or natural gas.

That scenario, as impractical as it may seem today, was seriously considered when Congress began drafting the 1990 Clean Air Act Amendments and, less than two years later, the Energy Policy Act of 1992.

After years of lobbying by oil, auto, and environmental interests, new energy and environmental laws do give alternative fuels a role not seen since the last energy crisis. However, the use of reformulated gasoline and other clean gasolines such as oxygenated fuel means most motorists won't have to trade in their gasoline-powered cars for another quarter century or so. Both alternative fuel proponents and government officials charged with implementing ambitious clean-fuel legislation point out that alternative fuels will only be used in niche markets today and will probably not take the place of petroleum in our lifetime. In areas such as California and some Northeast states, alternative fuels, including electric vehicles, are expected to play a greater role than in other areas of the country because of mandates.

Still, in general, alternative fuels won't be replacing significant volumes of petroleum for awhile. The Department of Energy, for example, says that by 2010, up to 620,000 barrels per day (b/d) of gasoline could be displaced by oxygenates and alternative fuels — a small number when one considers that U.S. motor fuel consumption is expected to be 12 million b/d.

Today's alternative fuel market may be small, but it is growing and could prove profitable in the long-term. Already, a small but significant market, including public and private fleets, must use alternative fuels in place of gasoline according to federal law. Therefore, it's not surprising that today's petroleum executive may be called on to supply a number of alternative fuels such as propane, CNG, or methanol to stay competitive. And choosing an alternative fuel can be tricky because of cost, infrastructure, and unresolved business issues such as insurance, quality control, and safety concerns.

With the changing marketplace in mind, this book is designed to satisfy several needs. Primarily, it is a reference book because it features accurate and descriptive information on basic fuel characteristics, economic issues such as availability and typical pricing strategies, and practical details on storage and distribution concerns. The book is also designed to be a self-learning tool for those managers new to the alternative fuel industry.

CHAPTER TWO

HISTORY OF ALTERNATIVE FUELS AND OXYGENATES

You'll never get people to sit over an explosion.

— Electric vehicle pioneer Col. A.A. Pope,
commenting on the gasoline automobile

It's been said that past is prologue. And this old adage stands true for understanding the current interest in alternative fuels. Back when German inventor Nikolaus Otto first produced the prototype of the modern combustion engine in 1877, petroleum was not a predetermined conclusion — many early automotive engines were powered by a variety of fuels including gaseous mixtures (natural gas, propane and hydrogen), steam, coal, and even manure. Twenty years later, it was a French-built electric vehicle that held the world land speed record of 65.75 miles per hour.[1] However, nearly all these early candidates were overshadowed by oil-derived fuels for several reasons. As a liquid, oil was easier to transport and refuel than steam or electricity, supply was plentiful, and the energy density of the fuel was adjustable for motorized vehicles.

Over the past century, policymakers and even individual motorists have rediscovered nonpetroleum fuels like ethanol, methanol, natural gas, and electricity to solve problems such as crude oil shortages, grain surpluses, or poor fuel quality. More recently, alternatives have reentered the public arena as an air pollution control strategy. This chapter outlines some of the past interest in a few of the more widely known fuels to help better understand their popularity today. Alternatives such as synfuels and solar power have not been included in the discussion because of the technical and economic barriers that currently limit their marketplace significance.

REFORMULATED GASOLINE

The oil industry in the late 1980s coined the term *reformulated* gasoline to describe a new generation of cleaner burning fuels which would meet the environmental standards posed under the Clean Air Act Amendments of 1990. But in reality, the oil industry has been reformulating and reinventing the way it makes gasoline and diesel fuel since the industry was born in the late 1880s.

When gasoline supplanted alcohol as the preferred motor fuel in the early part of the automotive age, fuel researchers discovered that antiknock compounds such as tetraethyl lead and alcohols could be used to improve gasoline quality by raising octane. Despite some scientists' health concerns about the antiknock gasoline additive, refiners preferred lead antiknock because it was less expensive and more widely available. In the 1930s, refiners saw the need to further increase gasoline octane and began working on a technology that became known as the fluid catalytic cracking (FCC) process.[2] FCC units improved crude conversion and helped boost fuel quality by producing high-octane gasoline through alkylating light olefins. During World War II, the Allies took full advantage of these new FCC-produced gasolines which allowed fighter planes to achieve their designed power output using 100 octane fuel.

After the war, demand for octane continued in order to satisfy Detroit's new generation of muscle cars which had higher compression ratios than cars made before World War II. The industry responded again: through the use of catalytic reforming, refiners produced high octane components like aromatics to move ahead in the octane race of the 1950s and 1960s.[3]

Another critical challenge occurred during the oil shocks of the 1970s. Skyrocketing crude oil prices and a shortage of light, easily-processed crude oil led to new technological developments in heavy crude oil processing. At the same time, refiners' anticipations of EPA's enforcement of lead phasedown helped to encourage both ethanol and methanol blend use during the late 1970s and early 1980s as a new, cleaner way to replenish the octane pool.

When crude oil prices subsided in the mid 1980s, alcohols no longer made economic sense, so refiners largely turned to aromatics to meet octane requirements. But lead phasedown turned out to be only a preview of what would be in store for the refining industry. Despite ongoing improvements in motor vehicle technology, government officials in the mid 1980s said car and truck pollution was still a major source of air pollution in the United States. To help bring urban areas into compliance with federal air quality standards, refiners were told in the late 1980s to lower the volatility of their fuel. Two years later, under the Clean Air Act Amendments of 1990, the industry would face its biggest challenge ever — to develop a whole new class of cleaner gasolines and diesel.

Not since lead phasedown have refiners faced so much capital investment. Oil companies had no choice but to spend tens of billions of dollars in investments to make clean fuel reformulations at a time when refining margins were already squeezed by low crude prices. Ironically, the industry's earlier efforts to meet EPA's lead phasedown requirements had helped exacerbate the current air pollution problem. As lead phasedown progressed, petroleum refiners had maintained octane quality by increasing the amount of aromatics which, in turn, led to more volatile and more pollution-forming fuels.[4]

Today, refiners have again adjusted their gasoline recipes, but the results have proven to be far more expensive than lead phasedown. Under federal clean fuel legislation, refiners and blenders must use oxygenates, including methyl tertiary butyl ether (MTBE), fuel ethanol, and ethyl tertiary butyl ether (ETBE), to meet the regulations. But once again, the industry appears to have solved a seemingly intractable problem, at least for now.

ALCOHOL FUELS

In the new era of reformulated gasoline, alcohol fuels are no longer touted as an environmental panacea. Nevertheless, policymakers continue to see a role for both ethanol and methanol, alone or as components in petroleum fuels, in future environmental and energy policies.

Over the past century, alcohol fuels have typically been marketed as gasoline extenders or replacements in times of high oil prices or large grain surpluses. Most recently, alcohol has reentered the public arena as an air pollution control strategy.

Alcohols were there at the beginning of automotive history, years before fossil fuels such as gasoline, natural gas, or propane were seriously considered. Early pioneers of the "horseless carriage" experimented with alcohol because it was a liquid fuel that was easier to transport or store than other fuels such as steam, hydrogen, or natural gas derived from coal.

The first modern combustion engine, the Otto cycle, premiered in 1877 and ran on alcohol, for example, and Henry Ford himself designed a Model T in 1908 to run on alcohol, proclaiming it the "fuel of the future." Ford and Otto also were drawn to alcohol fuels because of their octane quality and boiling ranges, which allowed for better combustion compared to locally-produced gasolines at that time. Early oil refiners in the late 19th century and early 20th century only had use of crude distillation techniques, allowing lighter fractions of crude and shale oil, but octane levels were still lacking.

At the turn of the century, most engine manufacturers marketed locomotive, tractor, and automobile engines powered by fuel alcohols. Competition between alcohol- and gasoline-powered vehicles was common, with races between cars staged at agricultural and engineering expositions all over Europe.[5]

While technologically superior to U.S.-refined petroleum at the time, the economics of alcohol stymied its growth then, as today, because corn-to-ethanol or wood-to-methanol distilling processes made the fuel twice as expensive as gasoline. Despite the unfavorable economics, Ford continued to encourage alcohol fuels, teaming with the Dow Chemical Corporation in 1935 to market an ethanol-gasoline blend called "Agrol" at 2,300 Midwest service stations. Eventually, the effort failed because ethanol suppliers, mainly farmers, could not compete with cheap petroleum. Inexpensive fossil fuels also led automotive engineers to concentrate on gasoline-powered engines even though this meant their engines had lower compression ratios. The octane-poor fuel of the time meant engineers had to cut back engine power in order to avoid engine knock which could severely damage the vehicle. Detroit's continued emphasis on gasoline eclipsed the use of alcohol as a transportation fuel for the next 50 years until the first energy crisis of the 1970s. Alcohols, especially ethanol, continued in the agricultural sector because of plentiful supply and grassroots support throughout the Farm Belt.

FUEL ETHANOL

Not until later in the 1970s, when oil prices rose dramatically and grain surpluses threatened the family farm's existence, did public interest grow in alcohol blends as a "new" way to extend and possibly replace petroleum with America's own abundance. Just as ethanol fuels derived from corn and other agricultural surpluses had first become popular in the Midwest, it was there that renewed interest began in the early 1970s, especially in Iowa, Kansas, and Nebraska. Technical concerns by automakers and supply considerations stalled ethanol's growth, even with high crude oil prices, until the government stepped in and provided subsidies designed to help the industry.

Photo 2-1   The automobiles of the mayor of Lincoln and the governor of Nebraska being fueled with an ethanol blend in Lincoln, Neb., April 1933. Courtesy of the State of Nebraska Ethanol Board

Buoyed by federal subsidies, fuel ethanol production skyrocketed in the late 1970s and early 1980s. With supplies less of an uncertainty, gasohol sales quickly accelerated. In 1978 only a few dozen Midwest stations offered gasohol, but by 1979 the fuel was available in more than 1,000 service stations. By mid-1980, more than 2,000 U.S. gasoline stations offered the product in 45 states. But even with all the new interest in alcohol-gasoline blends, gasohol represented less than 1 percent of total U.S. gasoline sales.

Almost as quickly as it captured America's interest, ethanol fuel again faded from view. Crude oil prices crashed in the early 1980s, making ethanol blends look uneconomi-

Photo 2-12 Customer pumping AGROL in Kansas, 1930s. Courtesy of the State of Nebraska Ethanol Board

cal even with government subsidies. For a time, ethanol remained extremely popular in Brazil, where ethanol made from sugar cane had been mandated at the pump to provide a market for cane farmers. But the government eventually followed the United States' lead when skyrocketing inflation and higher world sugar prices caused the country to scale back its own ambitious program dramatically to include only ethanol-blended gasoline.

In the United States, Midwest support for the fuel, combined with hefty state tax incentives, has historically guaranteed ethanol's future, at least regionally, even in times of record low gasoline prices. Today, clean air programs such as the EPA's oxy-fuel and reformulated gasoline programs have revitalized the industry to a level not seen since the oil crises of the 1970s.

FUEL METHANOL

Both the energy crisis and lead phasedown catapulted methanol fuels back into the public's interest after being almost abandoned in the early 1900s. The one exception was the racing car circuit, which continued to use alcohol because of its high octane quality.[7]

As with ethanol fuels, the oil crises of the 1970s led to a rebirth in the methanol fuel transportation field. With the memory of gasoline lines fresh and air pollution problems accelerating in the early 1980s, California policymakers were the first to seriously consider using methanol as a total, or "neat," replacement for gasoline. The state, perhaps more than any other part of the country, had been dramatically impacted by oil crises of the 1970s. The state has a huge appetite for motor fuel and is the third largest consumer of petroleum products in the world.

In 1981, the California Energy Commission (CEC) decided to expand its methanol vehicle program. CEC first started with blend tests of alcohols for a variety of motor vehicles. This eventually evolved into a large government demonstration program for dedicated M85 vehicles, which use 15 percent gasoline and 85 percent methanol.

In 1982, the commission attempted to establish a methanol fueling network to support the vehicles, and a modest infrastructure was established at 15 independent gasoline retail stations. While the stations were not always in convenient locations, the commission planned the system so a driver with careful planning could travel from northern to southern California in a dedicated methanol vehicle. Following the technical success of methanol prototype vehicles, the commission bought a fleet of 150 M85 Ford Escorts and placed them at a local car rental agency at the Los Angeles International Airport for use by state employees on business.[8]

But government and industry officials ran into consumer acceptance problems soon after because they failed to recognize the effect of unfamiliar technology on motorists. In theory, driving methanol-powered vehicles seemed to be a minor adjustment for a motorist to make.

Market surveys had shown that state employees usually went to Los Angeles on one-day business trips; they drove downtown and back to the airport without refueling. Therefore, program planners did not expect the state's limited methanol refueling network to pose a problem, since their methanol cars, even with their limited driving ranges, would probably not need to be filled up. In reality, the insecurity of driving a vehicle with new technology, combined with sparse refueling locations and worries of getting lost in a different city, led some drivers to simply abandon the vehicles. Within six months, the program was discontinued, and the vehicles were sent to government fleet operators knowledgeable about alternative fuels.

To alleviate these concerns, the state slowly expanded its network of methanol refueling stations to 54. CEC encouraged fuel supplier participation by agreeing to split the cost of new refueling sites. In 1987, the introduction of fuel flexible vehicles, which can run on any combination of gasoline and alcohol, helped policymakers sell the idea of neat alcohols to fleet managers who had reservations about fuel supply and the driving range of neat alcohol vehicles.

GASEOUS FUELS

U sing gaseous fuels such as compressed natural gas (CNG), liquefied petroleum gas (propane or LPG), and liquefied natural gas to power vehicles is not a new concept, although their use was never as widespread as alcohol or electricity.

Natural gas from coal and electricity were both experimented with as motor fuels when the internal combustion engine was being perfected in the 1880s. T. Phillips, still a major propane supplier, was the first to convert V-8 engines to liquefied petroleum gas in 1913 because of the fuel's superior octane qualities compared to gasoline. However, engineers eventually abandoned the concept of gaseous fuel vehicles because of handling issues and concerns over the compression ratio required. At the same time, natural gas and propane were being increasingly used in place of whale oil in residential areas to heat homes, cook meals, and provide indoor lighting. Seeking to make natural gas more portable, Godfrey Cabot developed a way to liquefy natural gas in 1914, but his patent did little to bolster natural gas vehicles (NGV) because petroleum derivatives were in much larger supply and less expensive.

It wasn't until decades later that NGVs and propane-powered engines would play a significant role, with the vast majority of vehicles operating outside the United States.[9] During the 1940s, researchers again touted the benefits of a fuel that could be used during war-created gasoline shortages. "Extensive road tests on ordinary internal combustion engines show the use of liquid methane as a transportation fuel to be a thoroughly practical proposition," wrote two English engineers in the *Journal of Institutional Fuel* in 1945.[10]

Like ethanol, propane use had its roots on family farms. Propane was important to U.S. agriculture since it was (and still is) used for crop drying, heating structures, and as a motor fuel. LPG-powered farm equipment was a Midwest presense before and after the gasoline fuel shortages of World War II because propane was often the only kind of fuel available in rural locations. After the war, propane carved a niche in urban areas for retail businesses using indoor vehicles such as fork lifts. Propane, unlike other petroleum-based fuels such as gasoline or diesel, eliminated the danger of carbon monoxide poisoning in poorly ventilated areas. Since the late 1940s and early 1950s, gasoline engines suffered from low power and low compression, and diesel fuel was not readily available, farmers continued to favor propane because it meant a bigger power boost.

So, they shaved the tractor's engine head to reduce the volume of the combustion chamber and switched to propane, which was already being stored at the farm for cooking and heating needs.[11] However, by 1960, the ascendency of heavy-duty diesel engines significantly reduced the use of propane as a transportation fuel although some fleets, particularly in the Southwest where propane was plentiful, continued to use the fuel because its clean burning properties extended engine life.

Outside the United States, both CNG and LPG have enjoyed some acceptance in a handful of countries because of the high cost of crude oil and/or the abundance of domestic fuel supplies. In Italy, for example, the government established a national policy of promoting propane and natural gas because of the scarcity of petroleum during World War II.

Historically, countries with relatively high alternative fuel market penetration such as New Zealand, Canada, and Italy realized those gains only after decades of providing government subsidies that allowed the fuel to compete at half the price of gasoline.[12] A good example is Canada. In 1980, a national energy plan was put in place to reduce oil imports and boost domestic energy. From 1981 to 1984, the government provided grants to consumers converting vehicles to run on propane. In 1983, a similar program was initiated for CNG. Currently, about 1 percent of Canada's total number of vehicles (140,000) use propane; 85 percent of these vehicles can operate only on propane, and 15 percent can operate on propane or gasoline. For CNG, there are about 30,000 vehicles on the country's roads; 97 percent are dual-fueled, and only 3 percent exclusively use CNG.

The program was successful because the government planned and cooperated with industry. Canadian officials moved forward with a two-phase program promoting propane first because it was already used in some vehicles and a fuel distribution network was already in place. Only later did officials phase in the CNG program by applying the lessons learned from their propane experience.

Today, Italy still has the most established network of NGVs and propane-powered vehicles in the world, with about 300,000 on its highways, due in large part to the prohibitive taxes placed on petroleum fuels. There are about 205,000 NGVs in the former Soviet Union, 105,000 in New Zealand, 100,000 in Argentina, 30,000 in the United States, and 170,000 in Canada.[13]

As with alcohol fuels, interest in gaseous fuels grew in the United States during the late 1970s as a result of the oil pricing shocks of the Arab embargo. Before 1984, NGV conversion equipment was not standardized to meet safety and certification standards, and this hampered the growth of an industry already facing the burden of promoting an entirely new refueling infrastructure on motorists. In areas where a refueling retail network existed, such as in the case of propane, or when a large fleet decided to invest in a fuel, as in liquefied natural gas at the San Diego Gas & Electric utility, the fuels made only modest gains.

Southern California Gas, for example, established a pre-oil crisis NGV program in an attempt to diversify its market base beyond the already saturated residential market.

Selling dual-fuel vehicles primarily to fleet operators, the program was moderately successful until the oil pricing crashes of the early 1980s. The program was finally discontinued in 1986, when fuel injection systems in cars made the engine conversions obsolete.

What has proved to be the biggest catalyst for NGVs is the advent of federal clean air laws that restrict or ban the use of gasoline in some cases. Before 1985, NGVs were converted from carbureted "open loop" gasoline systems. These systems were designed for idle and acceleration conditions, but did not allow the conversion system to adjust the air-fuel mixture efficiently enough to control exhaust pollution. With the advent of computerized fuel injection in the mid-1980s, conversion technology became much more sophisticated, using a closed-loop system working in tandem with the engine's computer, so engine performance did not suffer and tailpipe emissions fell dramatically.

ELECTRIC VEHICLES

At the turn of the century, electric vehicles (EV) outnumbered gasoline-powered vehicles. In the era before mass production, automobiles were motorized toys of wealthy urbanites, and consumers, particularly women, preferred the quiet ride of an EV to a noisy, sputtering exhaust-ridden trip in a gasoline car. Even Clara Ford, the wife of Henry Ford, owned and operated a battery-powered Detroit Electric Vehicle, according to *Ward's EV Special Report of 1993.*

Battery-powered vehicles existed as far back as 1839 but were powered by a nonrechargeable battery that limited their use to entrepreneurs and tinkerers. But with the advent of rechargeable lead-acid batteries, EVs were on the market by the early 1890s, although clearly only a select few could afford such an expensive novelty. Even the famous car designer Ferdinand Porsche designed a handful of EVs, including the first "hybrid" car that combined a gasoline engine with an electric motor.[14]

From the very beginning, the cost of an EV appeared to outweigh its many advantages. In October 1899, Professor Elihu Thomson of General Electric discounted the future of EVs, which were competing with steam- and gasoline-powered vehicles, calling the limited driving range "an unmitigated nuisance" and the operating cost too expensive.[15]

While Thomson may have accurately predicted EVs' place in history over the next century, he missed the mark about gasoline vehicles. Thomson predicted that steam would power the automobiles of the future. His reasoning was twofold — gasoline was too dangerous to transport and too expensive to rely on as an energy source. When he made this prediction, much of the world's oil reserves (including North America's) were still undiscovered, and the price of gasoline made from known crude reserves had doubled in three years. Also, many local municipalities banned the transport of gasoline on public roads, bridges, and ferries, further complicating its sale.

Less than 10 years later, EVs would be virtually shut out of the marketplace in favor of gasoline-powered internal combustion engines. There were three main reasons for this: (1) The discovery of crude oil in California and Texas brought gasoline prices to within reach of the average consumer, (2) The invention of the electric starter made it easier to drive gasoline-powered vehicles, and (3) Henry Ford decided to mass-produce internal combustion engines instead of electric.

As with nearly all alternative fuels, EVs have enjoyed periodic, limited rebirths in the automotive marketplace in response to environmental or energy crises. In the 1960s, Ford Motor Company, General Motors (GM), and the former American Motors Corporation experimented with EV prototypes for the California market along with foreign automakers such as BMW. In 1964, for example, GM's engineering staff produced the Electrovair I, a Chevrolet Corvair with a silver-zinc battery pack. Two years later, GM demonstrated the Electrovair II. According to GM, its Electrovair II rivaled the performance of a standard Corvair piston engine, but with a range of only 40 to 80 miles. There was very limited market potential, even with growing interest in air quality issues. The company continued with its electric car development through the 1960s and into the 1970s. Then came the energy crises of the 1970s, and the idea of an electric commuter car gained further interest. GM's Chevrolet division developed the Electrovette, based on its Chevette, powered by 20 12-volt lead-acid batteries similar to EV models 10 years later. With gasoline prices expected to climb to $2 per gallon, GM predicted that 10 percent of its production in 1990 could be EVs. But history repeated itself — gasoline prices fell and GM and its competitors curtailed their research again, this time for only a decade as concerns about the environment and global warming brought electric cars back onto Detroit's agenda.

Today, these same environmental concerns have led states such as California to mandate these vehicles in an effort to make EVs a market reality. With the help of increased federal research money, automakers, entrepreneurs, and battery companies are working to accelerate battery technology enough to make EVs more practical for their users. As a result, the range of EVs has increased substantially since even the 1960s, but automakers say a lot more needs to be done.

References

1. Cannon, James S. *Paving the Way to Natural Gas Vehicles* (Inform, Inc., New York, N.Y., 1993).
2. Winfield, M.D. "Meeting the Environmental Challenge with Flexible and Economic Refining Technology," The World Conference on Refinery Processing and Reformulated Gasolines, San Antonio, Texas, March 23-25, 1993.
3. Leffler, William. *Petroleum Refining for the NonTechnical Person*, (PennWell Books, 1985).
4. Owen, Keith and Coley, Trevor. *Automotive Fuels Handbook* (Society of Automotive Engineers, 400 Commonwealth Drive, Warrendale, Pa., 1990).
5. Kovarik, Bill. *Charles F. Kettering and the Development of Tetraethyl Lead in the Context of Alternative Fuel Technologies.* SAE Paper 941942, presented to the Fuels & Lubricants Meeting & Exposition, Baltimore, Md., October 1994.
6. Segal, Migdon. *Alcohol Fuels* (Briefing paper of the Congressional Research Service, Washington, D.C., May 1989).
7. Sanchez, Felix. "Methanol Fuels," Alternative Fuels Refueling Equipment Technical Conference and Trade Show, Orlando, Fla., March 16-17, 1994.
8. Sullivan, Cindy and Patterson, Sue. "Launching the Market for Alternative Fuels," *1994*

Windsor Workshop on Alternative Fuels, Toronto, Canada, June 13-15, 1994.

9. *Natural Gas Vehicles...The Decision Starts Here*, First Edition, RP Publishing. January 1994.

10. Branson, Jerrel. "Liquefied Natural Gas Systems," *Alternative Fuels and Clean Cities Conference*, Milwaukee, Wis., June 30, 1994.

11. Clark, William. "Let's Get Both Acts Together," *Butane-Propane News*, June 1994.

12. Rezendes, Victor. *Experiences of Brazil, Canada, and New Zealand in Using Alternative Motor Fuels* (GAO Report, Washington, D.C., May 1992).

13. Sperling, Daniel. *New Transportation Fuels: A Strategic Approach to Technological Change* (University of California Press, Berkeley, Calif., 1988).

14. Nelson, Walter Henry. *Small Wonder: The Amazing Story of the Volkswagen* (Little, Brown, 1970).

15. Perrin, Noel. *Life with an Electric Car* (Sierra Club Books, 1994).

CHAPTER THREE

CLEAN GASOLINES: REFORMULATED AND OXYGENATED FUELS

Clean fuel laws have led refiners to modify one out of every three gallons of gasoline they sell in the United States. During the winter, for example, many states require that fuel compounds containing high levels of oxygen be added to fuel to combat carbon monoxide pollution. In a separate program, the federal government mandates that nine metropolitan areas use a special reformulated gasoline year-round that is especially designed to lower volatile organic compounds and other pollutants which contribute to ground-level ozone (see Appendix I for specific details of these programs).

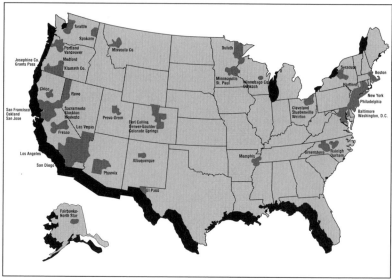

Fig. 3-1   Carbon Monoxide Nonattainment Areas   Source: U.S. Environmental Protection Agency

Other areas of the country have also voluntarily decided to join the federal reformulated gasoline (RFG) program because state regulators see clean fuels as a cost-effective way to comply with air pollution guidelines. In some parts of the country, the loss of the RFG program would result in more draconian measures such as odd/even driving days, mandatory carpooling, or even bans on barbecuing, according to EPA officials.

As a result, oxygenates, typically alcohol and ethers, play an important role for refiners, who must meet U.S. EPA oxygenated fuel and RFG requirements. An oxygenate is a fuel component added to gasoline or diesel to increase the amount of oxygen in the blend. EPA allows several oxygenates to be used in the gasoline pool, including fuel ethanol (C_2H_5OH), an alcohol which is primarily made from agricultural feedstocks, predominately corn; methanol (CH_3OH), another alcohol usually made from natural gas; and a variety of ethers such as MTBE ((CH_3)$_3COCH_3$) and ETBE ($C_2H_5OC_4H_9$) that can be produced by combining methanol, ethanol, or higher alcohols with a petrochemical feedstock called isobuytlene. Another class of oxygenates, known as esters, is being looked at in the United States and abroad as a way to reduce diesel emissions. Esters can be made from vegetable oil or alcohols but are not as commercially attractive as alcohols and ethers to today's refiner.

Fuel characteristics

Unlike other alternative fuels, oxygenated and reformulated fuels can be used in any car, truck, snowmobile, or lawnmower that uses conventional gasoline. For this reason, many states and the federal government consider RFG and oxy-fuels to be *replacement,* not *alternative,* fuels because many policymakers and consumers equate alternative as a noncrude source.

Both fuels developed in response to federal clean air laws designed to combat mobile source pollution from cars. Oxy-fuels are used in areas that do not meet federal carbon monoxide guidelines. Reformulated gasolines, which must contain an oxygenate under federal law, are designed to control ground-level ozone. Of the two programs, the oxy-fuel program is the most straightforward in terms of fuel formula and compliance.

Simply put, an oxygenated fuel is any gasoline, conventional or reformulated, that contains 2.7 percent oxygen by weight. Under federal clean air laws, refiners using conventional fuels in areas not required to use RFG still must take care the fuel is no dirtier than 1990 levels. Reformulated gasoline marks the most significant change to gasoline ever and the fifth time in 20 years refiners have taken steps, sometimes voluntarily and sometimes under government mandate, to produce cleaner burning fuels.[1]

EPA designed the RFG program to be implemented in two stages. Under Phase I, refiners were required, beginning January 1, 1995, to sell the so-called *simple* model which contains a specific fuel recipe through 1997 or to market a *complex* model, using a sophisticated computer program to determine the unique relationship between their fuel and regulated emissions. An EPA-approved model methodology gives each refinery the flexibility to meet emission standards by letting it vary certain individual parameters. In other words, refiners decide the best recipe for RFG without having to use EPA's cookbook.

However, from a practical standpoint, most refiners have opted to stick with the simple model during the 1995-1997 period for two reasons. First, EPA took such a long time to issue a final rule on its complex model that refiners did not have enough time to anticipate what investments would be needed to comply with the law.

Second, and perhaps more important, was the issue of convenience. Any complex-model RFG introduced into the market before 1998 must be kept in segregated storage, limiting the market value of the fuel since it cannot be commingled with other refiners' RFG.

Between 1995 and 1998, refiners are following a reformulated gasoline recipe that regulates five chemical properties of gasoline, including Reid vapor pressure, minimum and maximum oxygen content, and maximum benzene content. Heavy metals such as lead or manganese may not be used at all. In addition, refiners must ensure a minimum 15 percent reduction in ozone-forming compounds from the refiner's baseline[2]. This baseline is the average fuel compositions, i.e., volume level of aromatics, that the refiner produced in 1990. Refiners who decided not to calculate their own average follow an industry average called the *Statutory Baseline*. As a result, smaller independent refiners, who could not afford to make the expensive refinery improvements necessary to make RFG use their own 1990 baseline, and larger multinational companies that were able to invest early in RFG, technology are using the industry standard.

In 1998, refiners will no longer have an option. RFG must meet the complex model standard.[3] This expands the performance standards a refiner must meet to sell a government-approved RFG blend. Various parameters such as oxygenate type, sulfur content, olefin content, and fuel distillation are all part of the mix, but a refiner does not have to

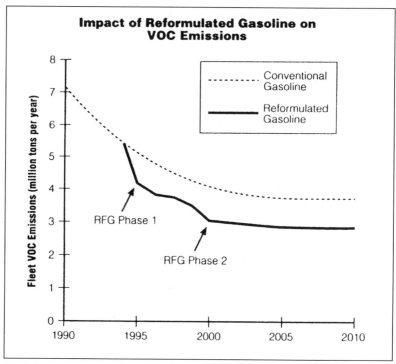

Fig.3-2   Impact of Reformulated Gasoline on VOC Emissions   Source: U.S. Environmental Protection Agency

meet a specific oxygen content (volatile organic compound, for example) to comply with the law. Rather, refiners must be able to show the fuel meets the 15 percent VOC reduction target. The complex model will also be enforced under Phase II of the RFG program, which begins on January 1, 2000, requiring a 20 percent reduction in toxic air pollutant emissions and a 5 percent reduction in nitrogen oxides when compared to a refiner's average gasoline composition in 1990.[4]

Depending on the severity of the air pollution problem, some ozone nonattainment areas that violate both carbon monoxide and ground-level ozone standards use both oxy-fuel and RFG programs to control emissions. To satisfy the needs of these programs, refiners sell a RFG gasoline with the higher 2.7 percent oxygenate content during at least four winter months. Then, at the end of the oxy-fuel season, the oxygen content of the RFG is reduced to 2.0% oxygenate.

STORAGE AND DISTRIBUTION ISSUES

Reformulated gasoline was first proposed by the oil industry in the late 1980s in an effort to stop U.S. lawmakers from banning gasoline in some of the nation's most polluted areas.[5] The rationale for reformulated gasoline was that both motorists and fuel suppliers could reap the benefits of clean transportation without having to dismantle a petroleum-based fuel distribution system worth tens of billions of dollars.

Refiners rely on a number of ways to transport petroleum products such as gasoline, diesel, or fuel oil to consumers. Fuels are delivered by pipeline, waterborne tanker and barges, rail, and tanker truck. Storage terminals also help to ensure smooth deliveries in the system by providing reserve supplies to meet unanticipated market conditions.

All this is lost on the typical motorist, however. Drivers don't care about how or why the gasoline got to the station. Gasoline merely provides the means to travel. But from a refiner's perspective, delivering RFG is a very complicated business. Since lead phasedown, product differentiation has increased within the motor gasoline business, and the government estimates there are more than 250 possible types of conventional, oxygenated, and reformulated gasolines.[6] Factors such as location, time of the year, and local laws all impact fuel composition almost on a daily basis for today's refiner.

As a result, new RFGs and their various grades are expected to further test the fungibility, or interchangeability, of the motor fuel business. As states have complied with the Clean Air Act, the nation is a checkerboard of areas requiring different types of gasoline at varying times of the year.

Because of these regulations, refiners say they are more vulnerable to sudden supply shortages brought on by natural disaster or other unforeseen market forces such as a refinery shutdown. With the advent of oxy-fuels and RFG, an increased number of products requiring segregation are delivered through a fixed capability pipeline system and then transferred and stored at a limited number of terminals. Especially sensitive to disruptions are rural locations using a single gasoline supplier and areas with limited terminal storage locations.

Further complicating the picture is the use of fuel ethanol, which is not pipeline fungible because it can bond with the water usually present on traditional pipelines. Ethanol requires different materials in delivery and storage facilities than ethers or traditional petroleum products need. Seals, gaskets, and tank liners all must contain materials that resist corrosion or swelling of rubberized parts. For this reason, neither ethanol-blended gasoline nor 100 percent neat ethanol can be shipped by pipeline, the least expensive and preferred means of shipment in the industry. Typically, fuel ethanol is shipped by railcar, truck, or barge, depending on its market destination. Conversely, MTBE is treated like any other petrochemical and can be distributed alone or in gasoline virtually anywhere gasoline travels.

TESTING

An important difference between conventional gasoline and clean gasolines is fuel quality testing. In the case of RFG and oxy-fuels, part of the EPA program involves testing and oversight measures to make sure everyone is following the rules from the refinery gate to the service station pump. Testing is not new to the gasoline industry. Since lead phasedown, federal and state policymakers have required some kind of testing, but never as widespread as today. Virtually every batch of RFG and conventional gasoline must be sampled and is subject to a variety of tests.[7]

What gets especially tricky for marketers is selling conventional gasoline near RFG areas since they have to ensure that their product doesn't get mingled. In the new era of reformulated fuel, fungibility becomes almost nonexistent because a product must be segregated according to the kind of oxygenate used, Petroleum Administration Defense District location, etc. As a result, it's not uncommon to see a major oil company with 16 or 20 different gasoline grades in one season. And lest one forget, not testing means risking a fine of $25,000 per violation per day, according to EPA.

Policies on testing and oversight vary with each gasoline supplier, but in the end, EPA will make a marketer share the blame for a violation with everyone along the distribution chain unless one has a good line of defense, such as a written contract.

Wholesalers and retailers also should have on file the three most recent product transfer documents of gasoline delivered to each tank retained at each retail outlet. They also must maintain the following records for five years: product transfer documentation; sampling and testing information that includes the location, date, time, storage tank, or truck identification for each sample collected; and the test results.

EPA also conducts its own compliance monitoring. About 160 weekly surveys are conducted at retail stations in RFG-covered areas by an industry contractor. Failure to conduct a proper survey could result in a tightening of standards for the entire marketing area, making compliance more difficult for all upstream and downstream parties.

To meet compliance guidelines, the majors send fuel samples to an outside, independent, EPA-approved laboratory, which confirms that the sample meets American Society for Testing and Materials requirements and the federal RFG program requirements (i.e., benzene levels, oxygenates, sulfur, etc.) Wholesalers and service stations can also use an outside lab to confirm fuel quality, but many use portable onsite equipment, which is slightly less accurate but provides faster readings and is less expensive.

HEALTH AND SAFETY ISSUES

Refiners treat today's clean gasolines, including oxygenated and reformulated gasoline, in the same fashion as they treat conventional gasoline from a public safety standpoint. As with gasoline, care should be taken to minimize skin contact and avoid breathing vapors or accidently ingesting fuel.[9]

The government, refiners, and environmental groups all have found reformulated gasoline to be as safe, if not safer, than conventional fuels from a public health standpoint. Nevertheless, EPA, in conjunction with the refining industry, has performed a comprehensive chemical inventory of pollutants produced by these fuels and related fuel additives, including oxygenates. The research will provide health effects data by the year 2000 on more than 7,000 fuels and fuel additives registered with EPA, including hundreds of reformulated gasoline types.

However, the relatively recent widespread introduction of oxygenates into the gasoline pool has caused some consumer perception problems. Prior to the start of federal clean fuel programs, few if any problems had been reported with oxygenates in the continental United States. In the Southwest, for example, seven cities had been using both MTBE and fuel ethanol to reduce carbon monoxide pollution for several years.[10] But in Alaska, the introduction of the oxygenate MTBE during the 1992 to 1993 season brought both high prices and a wave of concern over possible acute health side effects such as nausea and dizziness. A variety of oxygenated industry-funded studies found that Alaskan base gasoline originating from north slope refiners may have intensified the odor of MTBE blends because it has a different chemical composition from fuel used in the lower 48 states.[11] Studies also have shown that poor ventilation attributed to arctic conditions at retail stations may have aggravated the smell. The following year, motorists in New Jersey voiced similar concerns even though MTBE-blended fuels had been used to boost octane in that region for several years.

Responding to these concerns, the EPA cooperated in an MTBE health study funded by the petroleum industry to discover why motorists in Anchorage, Alaska, were suffering from the same types of headaches and nausea symptoms as motorists in northern New Jersey. During the early 1990s, additional studies were performed by independent laboratories, including one involving people with known sensitivities to very low-level chemical exposures and chronically ill individuals, which found no association between reported symptoms and exposures to MTBE.[12] EPA officials concluded that adding the oxygenate to gasoline did not pose a health risk to the general public. Since then, both government and industry officials say more than 100 published studies of the health effects of MTBE have shown no evidence of increased risk to human beings at levels associated with its use in either program. Government and industry toxicology experts concluded the high level of publicity surrounding these fuels, coupled with price increases, may have caused some to associate health complaints with oxygenated blends even though no scientific link could be established.

EPA, oxy-fuel groups, and refiners maintain that reformulated gasolines, which often contain MTBE, may actually be less carcinogenic than conventional fuel because tailpipe emissions of air toxics drop by 20 to 22 percent. For example, by using MTBE blends, emissions of benzene, a known carcinogen, fall as much as 24 percent, according to the EPA.

ECONOMICS

In general, clean gasolines are less expensive than alternative transportation fuels and provide greater convenience to motorists. In research done by the American Petroleum Institute (API), either the price of RFG itself is less expensive or the overall costs of a fuel, vehicle, and fuel delivery network are less. For example, API found that RFG is about half as expensive to produce as some alcohol fuels. And reformulated gasoline is less expensive than electric battery power or compressed natural gas because it does not need costly vehicle

U.S. REFINING CAPITAL EXPENDITURES (Billions of 1990 Dollars)			
	1981-1990	1991-2000	2001-2010
Environmental, Health, and Safety Stationary Source Facilities			
Pollution Abatement – Reported*	6.1	—	—
Stationary Facilities – NPC Estimates	3.9	22.8	13.5
Subtotal Stationary Source Facilities	**10.0**	**22.8**	**13.5**
Refinery Process Facilities			
Product Quality	†	13.7	NS
Process Additions & Replacement	37.5	18.3	15.0
Total Capital Expenditures	**47.5**	**54.8**	**28.5**
Memo:			
Total Regulatory Compliance Expenditure	N/A	36.5	13.5
% Stationary Source Facilities	21%	42%	47%
Average Refining Capacity, Million Barrels per Stream Day	16.7	16.5	16.5
Capital Expenditures, $/Daily Barrel of Capacity	2,840	3,320	1,730

* From U.S. Department of Commerce, MA-200.

† 1981-1990 product quality facilities included in process additions and replacement.

Note: NS = Not Studied. N/A = Not Available.

Table 3-1   U.S. Refining Capital Expenditures. Source: National Petroleum Council

modifications or a new refueling infrastructure.

While it is still less costly than some of the alternatives, RFG still costs more to make than conventional gasoline for a number of reasons. Refiners must substitute higher priced oxygenates for typical gasoline components such as benzene, butane, and pentane; they must test fuel samples and keep records to meet EPA rules; and they have been forced to retool their refineries and modify their storage and distribution systems. All told, both government and private industry say that oil companies have invested $14 billion already to make cleaner fuels and more than $50 billion will be spent before the end of the decade.

All this investment has meant slightly higher prices for consumers — anywhere from three to eight cents per gallon, depending on locality and general market conditions such as supply and demand.

OXYGENATE SUPPLY AND DEMAND

Refiners' attitude toward oxygenates is a love-hate relationship. All oxygenates reduce emissions and add octane, but they also reduce gasoline throughput and cost more than traditional fuel components available at the refinery. Still, oxygenates are the lesser of two evils in the minds of many refiners. Without oxygenates, today's gasolines could not meet clean air regulations and would be supplanted by cleaner-burning alternatives provided by natural gas and electric utilities that are ready, willing, and able to replace the gasoline pump with an outlet socket or fast-fill clamp.

Oxygenate production facilities are located near their feedstock source.[13] Most etherification facilities are located on the Gulf Coast to take advantage of the area's vast supply of natural gas and petrochemical supply. On the other hand, about 80 percent of the nation's fuel ethanol facilities are located in the heart of the Midwest Corn Belt. Now and into the future, refiners can rely on ample feedstock supplies which will meet the demand for oxygenates. Also, new advances in catalytic cracking, a gasoline manufacturing process, have expanded the refiner's ability to produce these ether building blocks.

MTBE AND ETBE

MTBE is the most predominate oxygenate in nearly every region of the country except the Midwest, where fuel ethanol plays a large role, garnishing up to 80 percent marketshare in some areas because of logistical considerations. Use of ETBE, the chemical sister of MTBE, which has an ethanol feedstock instead of methanol, is expected to grow later in the decade when refiners are under stricter clean fuel guidelines that require their fuel to be even less reactive. ETBE is typically more expensive than MTBE but is much less reactive. Today, many refiners that can produce MTBE themselves have retrofitted their operations to accommodate either methanol or ethanol so they can adjust to changing market conditions.

MTBE is produced by reacting methanol with isobutylene using an ion-exchange resin catalyst. Although there are several different types of MTBE process techniques available, all depend on the source of isobutylene. Isobutylene is derived from four main sources:

■ raffinate streams coproduced with ethylene in steam crackers

■ refinery FCC butane-butylene streams

■ dehydrogenated-tertiary butyl alcohol from propylene oxide plants

■ field butane using both isomerization and dehydrogenation processes

Use of other ethers, such as tertiary amyl methyl ether (TAME) and tertiary amyl ethyl ether (TAEE), has also expanded as refiners have reconfigured their refineries to produce more isoamylene feedstocks. However, their use is limited to individual refinery configurations.

To meet clean fuel requirements, scores of gasoline refineries worldwide built or added small to medium 1,000-3,000 barrels per day, ether units. Adding small, captive ether supply is the most economical way to make oxygenates for refiners, although world-class plants in Saudi Arabia, Canada, and other areas have good access to inexpensive field butanes. However, even with captive ether units, most oil companies also purchase MTBE, ETBE, and ethanol on the merchant market because their own production is limited to their isobutylene output.

The cost of MTBE depends on several factors, including plant location and size, labor, and the source of feedstocks such as methanol and isobutylene. In general, producing MTBE from refinery byproduct isobutylene is the least expensive method. Stand alone MTBE units that use field butane or isobutane to make an isobutylene feedstock prior to MTBE production are more expensive because of the high capital cost to convert the raw materials into isobutylene. Of course there are always exceptions. When world methanol and butane prices are depressed, stand-alone plants can be very competitive, even when factoring in the cost of transportation, whether it be from the U.S. Gulf Coast or the Persian Gulf.

When clean fuel programs were first announced by EPA, there were concerns in the refining industry that there would not be enough oxygenates to meet increased demand. Luckily, these fears were never realized. Production from MTBE and ethanol plants in the United States can meet the vast majority of demand for clean fuel programs. Imports play a small but significant role since they keep prices competitive and offer an alternate form of supply in the event of a plant closing or unusually high demand.

Also, it is important to note that while MTBE and fuel ethanol are used primarily for their clean air benefits in the United States, in other nations, particularly those in Europe, they are used mainly in the lower-priced octane markets and compete with traditional petrochemicals such as benzene, toluene, and xylene.

FUEL ETHANOL

Fuel ethanol is made mainly from corn, and production is dominated by a handful of large agriprocessors that also make other products such as corn sweeteners and corn oil. For the most part, ethanol production would be extremely unprofitable, except the U.S. government provides tax incentives to encourage a domestic industry. Many Midwest states also have additional subsidy programs that make it economically attractive to produce and sell ethanol fuels. Two factors limit ethanol's use nationwide. First, ethanol blends cannot be shipped by pipeline, and therefore, direct blending often occurs outside the refinery gate. Second, the ethanol source should be as close as possible to where the blending takes place to hold down on transportation costs.[14] Expanded fuel ethanol production also depends on the availability of corn, which is primarily grown for food. While fuel ethanol production can look economically attractive during crop surpluses, the possibilities of drought or expanded demand for food products have dissuaded refiners from relying too heavily on fuel ethanol for oxygenate supply.

In the United States, two different processes — dry milling and wet milling — are used to produce ethanol from the starch in corn. Today, about two-thirds of those plants use the more sophisticated wet mill technique, which allows an agriprocessor to make a wide variety of products, including ethanol, corn sweeteners, corn oil, and animal feeds. The cost of producing ethanol varies widely from plant to plant, depending on the original decisions concerning the plant's location, process, size, energy source, and whether it stands alone (see Chapter Six.)

Inherently, ethanol's major value should be as a result of its high blending octane number and its oxygen content, which reduces carbon monoxide in tailpipe emissions. Before the advent of clean gasolines, ethanol was confined to downstream blending with the resulting punitive pricing

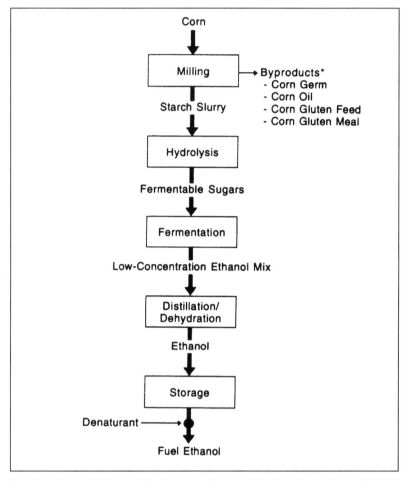

Fig. 3-3 Steps in Ethanol Production. Source: Paul R. Wood, "New Ethanol Process Technology Reduces Capital and Operating Costs for Ethanol Producing Facilities," Fuel Reformulation (July/August 1993) pp. 56-64.

structures, and its use was mainly limited to the Midwest, where the vast majority of production is located. Today, it has gained slightly wider acceptance in some markets outside the Midwest, but most of its growth potential is being redirected to other oxygenates such as ETBE, which is not as reactive and can be blended in refineries and transported by pipeline.

ENVIRONMENTAL

Oxygenated fuels, including reformulated gasoline, were developed specifically because of environmental concerns with conventional gasoline. Just as removing lead from gasoline eliminated the threat of lead poisoning, cleaner gasolines produce less of the pollutants that create smog and a variety of carcinogenic air toxics such as benzene.[15] By mandating the use of these fuels, policymakers have sought to limit the public's exposure to mobile source air pollutants which have been linked to a number of health conditions, including asthma and lung cancer. Smog can especially be a problem during the summer because ozone tends to form when heat and sunlight combine with auto emissions.

Adding oxygen to gasoline improves fuel combustion, resulting in less harmful tailpipe emissions. That's because oxygenates burn cleaner and more completely than do chemically complex, ring chain hydrocarbons like aromatics. This more efficient burn results in a fuel that contains 15 percent less carbon monoxide tailpipe fumes than conventional gasoline.

Reformulated gasoline is an even more effective air quality tool than traditional oxy-fuels because it reduces ground-level ozone and air toxics along with carbon monoxide. Policymakers estimate that nearly 1.3 million tons of ozone-forming emissions will be prevented in the first phase of the RFG program, 1995 to 1999, and that reductions will be even greater during the second phase of the EPA program. Total air toxics, including the carcinogenic compound benzene, drop 22 percent when using reformulated fuels. RFG also reduces particulate and hydrocarbon pollution by 15 and 10 percent, respectively.

GREENHOUSE GAS EMISSIONS

Mobile sources make up about 30 percent of greenhouse gas emissions, which either directly or indirectly cause global temperatures to change. The most common greenhouse gases are carbon dioxide, water vapor, methane, nitrous oxide, and carbon monoxide. In burning petroleum-based fuels, including RFG and oxy-fuels, it is thought that through the combustion process, the carbon dioxide released may over-encourage heat to be retained around the Earth the same way humid air inside a greenhouse is trapped. So far, only limited research has been performed in this area, but preliminary government estimates suggest replacement fuels like RFG emit less carbon dioxide than conventional gasoline and thus have a lower global warming tendancy. In the case of some oxygenates, such as ethanol, however, some emissions may be greater than gasoline if one considers the total fuel cycle from wellhead (or in this case corn stalk) to tailpipe.

CONCLUSION

Ongoing improvements in both refining and automotive technology ensure that for the vast majority of motorists, reformulated gasoline will be the one and only alternative fuel in place for the foreseeable future. In fact, refiners say the next generation of reformulated fuels, used in concert with evolving automotive technology such as lean-burn engines and pre-heated catalysts, may also achieve the same emissions reduction as natural gas, the cleanest burning fuel next to zero emission electric vehicles. Other replacement fuels for conventional gasoline, such as low-Reid vapor pressure (RVP) fuel, may supplant RFG in areas that choose not to use oxygenated fuels because of price considerations, refiners predict. However, the air quality track record of these fuels remains uncertain at this writing.

References

1. American Petroleum Institute. *Reformulated Gasoline Costs More*, Editorial and Special Issues Department, November 17, 1994.
2. Osborne, Tom. "Reformulated Gasoline Primer," *Independent Gasoline Marketing Magazine*, August/September 1994.
3. National Petroleum Council. *Meeting Requirements for Cleaner Fuels and Refineries*, Volume I, Analyses and Results, August 1993.
4. Energy Information Administration. *Assessment of Reformulated Gasoline*, Volume I, October 1994.
5. Lorenzetti, Maureen. "Passing the RFG Test," *Independent Gasoline Marketing Magazine*, December/January 1995.
6. Energy Information Administration. *Assessment of Reformulated Gasoline*, Volume I, October 1994.
7. Lorenzetti, Maureen. "API: Reformulated Gasoline May Be Cleaner Than Methanol," *Oxy-Fuel News*, August 14, 1989.

8. Lorenzetti, Maureen: "Regulation and the Environment," *Platt's Oilgram News*, September 12, 1994.

9. Owen, Keith and Coley, Trevor. *Automotive Fuels Handbook* (Society of Automotive Engineers, 400 Commonwealth Drive, Warrendale, Penn., 1990).

10. Schafer, Shannon. "OFA-Funded Study Provides Reasons for Alaska MTBE Scare," *Oxy-Fuel News*, August 29, 1994.

11. Oxygenated Fuels Association. *Scientists Give MTBE Clean Bill of Health*, Press Statement, January 18, 1995.

12. Department of Energy. *Alternative Fuels Feasibility Study*, Final Draft, December 9, 1994.

13. Oxygenated Fuels Association. *Benefits of a National Oxygenated Fuels Policy*, White Paper, 1994.

14. Ludlow, Bill. "Finding the Best Role for Ethanol—Revisited," *MTBE, Oxygenates, New Fuels Newsletter*, February 16, 1995.

15. Environmental Protection Agency. *Reformulated Gasoline: A Major Step Toward Cleaner Air*, Pamphlet, June 1994.

CHAPTER FOUR

CLEAN DIESELS: LOW SULFUR AND SOYDIESEL

Although diesel fuel may only represent 16 percent of the petroleum burned in U.S. engines, one cannot discount its role in the transportation fuels market. It is the fuel of choice for many heavy-duty vehicle applications, including freight trucks, transit bus service, and off-road construction equipment. In the United States, the population of heavy-duty trucks is small in comparison to automobiles and light-duty trucks, but the vehicle miles travelled is significantly higher. According to U.S. government figures, more than 20 billion gallons of diesel fuel are used each year in the United States for on-highway purposes, with heavy-duty trucks and transit buses taking up most of this amount. Another 3 billion gallons are burned in farm use while 2 billion is sold for military and off-road applications. Diesel fuel demand is expected to climb as highway freight traffic rises.[1] With demand for this fuel growing, alternatives to diesel fuel are assured a small but significant role for heavy-duty engine fleet managers who must comply with the Clean Air Act's Clean Fuel Fleet regulations. For most diesel users, and the vast majority of the motoring public, reformulated or low-sulfur diesel will be the fuel of choice because of price and supply considerations. New lower standards for sulfur content of on-highway diesel fuel took effect October 1, 1993 nationwide and currently affect half of total U.S. distillate demand, or 8.2 percent of total U.S. petroleum product demand. California has its own stricter diesel standard as well.[2]

Along with more conventional clean diesel fuels, there are some small niche markets where clean burning additives or substitutes called "biodiesel" are available that require little or no engine modification. Supporters of these fuels see their product as a way to encourage agricultural markets and further reduce air pollution, similar to the role fuel ethanol plays. Biodiesels are typically derived from the fats and oils of plants or animals, and can't compete in price with traditional or reformulated diesel fuels except in those Midwest locations that offer specific subsidies for their use. Other alternative fuels, including CNG and LPG, also can be used in diesel engines but require more extensive modifications. The following chapter focuses specifically on low-sulfur diesel and biodiesel fuel. For more information on other diesel engine alternatives fuel application, refer to Chapter 11.

FUEL CHARACTERISTICS

BIODIESEL

Biodiesel is a broad name that refers to both additives and substitutes for diesel that are based on methyl esters of vegetable oil or fats. Like ethanol, it can be produced from either a petrochemical feedstock or biomass such as soybean or rapeseed. One kind of diesel alternative, for example, is soydiesel, made from soybean oil. As a substitute for conventional diesel fuel, neat biodiesel is a clean-burning, non-toxic, biodegradable fuel, but the fuel is more popular as an additive to conventional or low-sulfur diesel fuels as a way to control emissions.

When made from a biomass feedstock, typically soybeans in the United States and rapeseed in Europe, biodiesel is made by a catalytic chemical process called *trans-esterification*, using an alcohol such as methanol and a catalyst. Methanol is mixed with sodium hydroxide and then with soybean oil, letting the glycerine that is formed settle. This forms fatty esters which are then separated into two phases, allowing easy removal of in the first phase. The remaining alcohol/ester mixture, methyl soyate, is then separated and the excess alcohol is recycled. The ester is then purified through a number of techniques, including water washing, vacuum drying, and filtration. Biodiesel can also be produced through a petrochemical route, but as with ethanol, this is rarely done unless the endproduct is for industrial or pharmaceutical purposes which demand extremely high purity levels. The Fischer-Tropsch Process, which synthesizes hydrocarbons from carbon monoxide and hydrogen, can be used to produce a wide variety of transportation fuels, including ethanol, methanol, gasoline, diesel, and biodiesel.[3] At the refinery, the gas can be supplied from a number of different sources, including coke or natural gas, by steam reforming or partial oxidation, but must contain hydrogen and carbon monoxide in a ratio of about two to one. The reaction is shown below.

$$nCO - 2nH_2 — CnH_2n - nH_2O$$
$$nCO - (2n-1)H_2 — CnH_2n-2 -nH_2O$$

The final product, whether made from a petrochemical or agricultural feedstock, closely resembles conventional diesel fuel, with a higher cetane number. The cetane number rates its starting ability and antiknock properties. Energy content, viscosity, and phase changes are similar to petroleum-based diesel fuel, but not identical. Biodiesel has a higher cloud point, which is the point at which a fuel begins to develop wax crystals during times of cold temperature, and higher viscosity. The fuel is typically blended at the 20 percent volume level along with 80 percent low-sulfur diesel fuel. By itself, the fuel is essentially sulfur free, emits significantly less particulates, smoke, hydrocarbons, and carbon monoxide. Nitrogen oxide emissions (NO_x) are similar to diesel in both neat and additive form. Biodiesel has a high flash point and very low toxicity if digested.[4]

LOW-SULFUR DIESEL

Low-sulfur diesel, like reformulated gasoline, was designed as a way to reduce tailpipe emissions without forcing motorists to invest in a whole new transportation fuel infrastructure. As with traditional diesel, low-sulfur diesel has a very low Reid Vapor Pressure, about 0.2 versus 8 to 15 for gasoline, which is consistent with the high boiling range of the fuel, between 370 and 650 degrees Fahrenheit versus 80 to 437 degrees Fahrenheit for gasoline. The cetane number of diesel is similar to the octane level of gasoline in that both are measures of ignition quality in an engine. Typical cetane levels in the United States run from 40 to 45. In Europe, where diesel is more prevalent, diesel cetane levels are 5 to 10 points higher, although U.S. clean air rules are expected to bring cetane quality more in line with European standards.

Before clean air laws were on the books, high demands for gasoline in the United States shifted the refinery focus to large cracking operations that left relatively low cetane blend components available for diesel fuel. Diesel fuel quality was, in effect, controlled by the crude oil mix and downstream processing to meet gasoline pool specifications. Refiners didn't have total control over fuel quality because ASTM guidelines and pipeline specifications set standards also, but the rules were broad enough to give refiners much more latitude over diesel properties than for gasoline. As a result, diesel fuel and No. 2 distillate fuel often left the refinery gate with more sulfur, lower cetane, or more aromatics based on changes in processing to maximize gasoline yield and quality.

But clean air regulation on the state and federal level changed this approach, with refiners required to make diesel fuels that control sulfur dioxide emissions and particulates. Today, the federal 0.05 percent sulfur limit and the 40 minimum cetane index limit are helping to eliminate highly aromatic fuels, which will improve overall fuel quality and lower emissions. Moreover, the well-publicized fuel quality problems first associated with low-sulfur, low-aromatics diesel fuel experienced in California in the early 1990s have been resolved, mainly through refinery adjustments and the addition of diesel ignition improvers. Fuel additives for deposit modification, cold flow improvement, and anti-wear normally present in traditional diesel are even more important for low-sulfur distillates which may impact O-rings in older trucks.

Storage and distribution

BIODIESEL

One of the biggest advantages of biodiesel alone or as a diesel blend is that it requires little to no modifications to fuel systems and supply networks. This gives biodiesel a significant advantage over other diesel engine alternatives, such as CNG or LPG, which cannot use the existing infrastructure. Biodiesel fuels require little additional capital investment in the refueling network, but there are three areas fuel handlers should be aware of when considering the use of biodiesel. These include material compatibility, cold temperature tolerances, and shelf life.

MATERIAL COMPATIBILITY

For neat blends, vehicle fuel lines and other components that come in contact with the fuel need to be changed because biodiesel can dissolve neoprene rubber hoses. Plastic hoses with brass fittings will solve any potential problems. Diesel fuel storage dispensing and fuel storage systems that feature elastomers made from butadiene, isoprene, hypalone, silicon, and polysulphide may also be vulnerable, and fuel suppliers acknowledge more testing on these surfaces needs to be done before the fuel is available on a widespread basis.

However, biodiesel blends in concentrations of 20 to 30 percent are not expected to cause material incompatibility problems and can be handled similarly to conventional diesel fuels. Fleet operators that have on-site refueling can choose to purchase biodiesel blends pre-mixed or they can have the biodiesel delivered directly to be blended with petroleum diesel on-site. No special equipment is needed to blend the fuel on-site, although a separate storage tank for the fuel may be desired to prepare limited batches of blended fuel.

IMPACT OF COLD TEMPERATURE

Biodiesel, alone or blended, has a greater tendency to cloud and stop flowing in lower temperatures than conventional or low-sulfur diesel, so fuel-heating systems or cold weather diesel additives are needed in lower temperature climates. Neat methyl soyate begins to cloud at 20 degrees Fahrenheit compared to No. 2 diesel's cloud point of -6 degrees Fahrenheit. A neat biodiesel fuel will begin to gel at around -20 degrees Fahrenheit. This property poses problems in northern climates since fuels should be rated at about 11 degrees Fahrenheit above ambient temperatures.

SHELF LIFE

Since neat biodiesel is usually produced from plant oils or animal fats, it could have less of a useable shelf life than traditional motor fuels because of the possibility of microbial contamination. Due to the limited production for the fuel at the present time, suppliers have not been able to provide a concrete estimate on how long biodiesel can be stored before fuel quality is degraded.

LOW-SULFUR DIESEL

Low-sulfur diesel follows the same distribution network as gasoline until delivery at the retail level, where the fuel is segregated (see RFG Chapter). In the United States, the bulk of retail sales are with heavy-duty commercial vehicles because of the fuel-efficient nature of the diesel engine, which is about 25 percent more efficient than gasoline engines. Diesel engines also tend to be more durable, tolerating more vehicle miles travelled. But the high initial cost of diesel engines tends to discourage their use in light-duty applications such as the passenger car market, in which fuel economy gains won't pay for the initial capital investment, except in areas where fuel costs are high.

For these reasons, three-fourths of the on-road diesel sold at retail is at truck stops or at company on-site refueling locations.

SAFETY AND ENVIRONMENTAL ISSUES

Biodiesel, when used as a replacement for diesel, has fewer safety and environmental risks associated with it than conventional petroleum fuels since it is nontoxic and biodegradable. When blended with diesel fuel, however, the finished blend should be handled like conventional diesel fuel. Similarly, low-sulfur fuels, from an environmental and public health standpoint, closely mirror traditional diesel fuels.

FLAMMABILITY

In the event of fire, diesel fire fighting techniques may be used for neat biodiesel and blended fuel fires. Fuel handlers, therefore, should be aware of the flammability of diesel fuels, including biodiesels. Flashpoint indicates the limiting temperature at which a fuel can be handled without danger of fire. At the flashpoint, fuel vapors form above the fuel surface in sufficient quantities for ignition to occur in the presence of an open flame. Diesel fuel and biodiesel have the highest flashpoint of any available transportation fuel, which translates to a higher level of safety compared to gasoline, alcohol, or gaseous fuels.

SPILLS

Biodiesel is completely biodegradable and poses significantly less environmental spill hazard than petroleum-based fuels. In the event of a biodiesel spill, fuel handlers should dilute the area with water only. The fuel will naturally disperse into its surroundings quickly. In the case of concrete, long-term exposure to methyl esters may weaken and crack concrete foundations, so it is important that continual spills in fuel handling areas be avoided.

ECONOMICS

SUPPLY AND DEMAND

Although biodiesel is gaining in popularity in Europe, little biodiesel is made in the United States even though potential capacity is large. U.S. soybean growers say there is as much as a billion pounds of surplus soy oil, capable of producing 133 million gallons of biodiesel. Another large untapped source is 2 billion pounds of yellow grease from used cooking oils and animal fats produced annually. This translates to 266 million gallons of biodiesel that could recycle waste generated from fast food restaurants over America.

Also, about 50 million acres of land idled by U.S. commodity programs could be cultivated to generate an additional 3 billion gallons per year of biodiesel and not interfere with current oil needs in the food and industrial sector. All told, about 3.4 billion gallons of biodiesel could be produced per year, if demand warranted.

But for now, demand is limited to various demonstration fleets, mainly because of price. The cost of the fuel is determined by the feedstock being used but is estimated to be $2.50 to $6 a gallon because of small-scale production and feedstock costs. This is nearly double the cost of diesel fuel. The U.S. Departments of Energy and Agriculture have estimated that large-scale production using today's technology could reduce biodiesel costs to $1.50 to $1.60 a gallon. But large-scale production is a long way off, unless crude oil prices climb to $35 or $40 a barrel. In the meantime, to offset the price differential, supporters of biodiesel have successfully convinced a number of Midwest states to extend tax breaks at the pump and at the production level that are comparable with ethanol. Attention on the federal level so far has been fairly muted.

Another factor to consider when examining the cost of biodiesel is that as a liquid fuel there is no need to retrain employees to pump fuel. Because conversion to biodiesel alone or in a blend requires no modification of existing infrastructure, a fleet operator can continue to use his existing engines and vehicles, which could become a significant cost, given the capital costs involved with gaseous fuels. This can be an important selling factor for urban transit buses, for example, where individual engines can cost tens of thousands of dollars to retrofit. Taking these economics into account, and given that a large percentage of buses use diesel-powered engines, biodiesel demonstrations today focus heavily on the urban transit bus market.

For biodiesel to gain nationwide acceptance, ways must be found to lower the fuel's production cost. Because of growing conditions and agricultural policy considerations, research to lower the cost of soy-based diesel is taking place in the United States, while European countries have been focusing on rapeseed derived biodiesel. Other possible feedstocks for biodiesel include bio-oils from corn, cottonseed, peanut, sunflower, canola, and rendered tallow, which comes from animal fat. The National Renewable Energy Laboratory is testing aquatic plants, such as microalgae, for possible oil production.

SUPPLY AND DEMAND

LOW-SULFUR DIESEL

When the federal government established low-sulfur diesel fuel nationwide in all but a handful of off-road applications, refiners adjusted processing and/or installed new equipment to handle the October 1, 1993 deadline. At the start of the program, however, there were major growing pains, which resulted from inexperience from both refiners and regulators on how the new fuel would impact older engines' fuel pump seals that required the lubrication associated with older fuel mixes. There were thousands of reports of failed engines, particularly in California, which was requiring a more severe fuel standard with low sulfur and low aromatic, and in colder locations where some fuel suppliers misjudged what kind of additive packages would be needed in severe climates. The result was a technical and political problem that led to fuel prices 10 percent higher than the previous year and spot shortages. Six months later, regulators on the federal and state levels found that severe hydrotreating of diesel fuel to reduce sulfur to the 500 parts per million standard may have been the primary culprit affecting O-rings and other rubber engine parts. Since then, refiners have moved to adjust their hydroprocessing units and also have learned to anticipate inventory problems sooner.

The experience, while painful, taught both the regulators and the regulated a valuable lesson. Any change in a conventional fuel requires long lead times, an informed public, and a flexible government structure to work with industry to achieve the desired public policy benefits.

With most refiners geared to the federal low sulfur standards, the biggest economic consideration for the industry is demand.

According to long-term government demand scenarios, motor gasoline consumption will increase by 0.7 percent annually between 1993 and 2010. Gasoline's share of transportation energy use will decline as the overall fuel efficiency of conventional light-duty vehicles continues to improve and sales of alternative-fuel vehicles increase. In 2010, alternative fuels are expected to displace some 465 thousand barrels of oil per day.[5]

Demand for diesel fuels, since they are used mostly to transport freight, will follow trends in industrial output. Distillate use is expected to increase by an average of 2 percent a year between 1993 and 2010, with low-sulfur diesels representing well over 95 percent of the total demand by 2005.

OTHER FUEL ISSUES

At this writing, EPA was not expected to mandate reformulated diesel fuel under a proposed rule that will be issued in late 1995 to curb heavy-duty engine emissions, agency officials said. Engine manufacturers maintain that both the petroleum and auto industry should share the regulatory burden of reducing mobile source emissions, similar to what is now being done for the light-duty vehicle market with reformulated gasoline. But the oil industry says it has invested all it can afford in clean fuel technology, including low-sulfur diesel and RFG, and that any remaining gains in air pollution control should come from Detroit, not the oilpatch.

EPA's rule is designed to drastically reduce emissions from new trucks and buses starting in 2004. The proposal would cut the NO_x standard in half to reach 2.0 grams/brake-horse-power-hour (g/bhp-hr) and retain the current 0.1 g/bhp-hr particulate standard for heavy-duty engines that run on gasoline or diesel. Agency officials say their action will "harmonize" federal and California heavy-duty vehicle programs, but it will not require stricter fuel standards.

So for the moment, the EPA appears to agree with fuel suppliers, although to keep the issue of reformulated diesel alive, EPA has agreed to continue research on developing cleaner engines and fuel technologies. With EPA's position on clean diesel fairly certain, heavy-duty engine makers like Cummins and Detroit Diesel say they are already adjusting to the new rules. Manufacturers have already developed some engines that run on alternative fuels, such as natural gas and methanol, in case diesel fuels cannot meet EPA's performance rules. Engine builders are also readying to meet mobile off-highway emissions standards. Cummins has unveiled new diesel-powered off-road engines that will perform at levels already adopted by EPA and the California Air Resources Board. New U.S. and California off-highway regulations will become effective starting January 1, 1996.[6]

ENVIRONMENTAL

Tailpipe emissions of diesel fuel are a source of mobile source air pollution but differ from gasoline emissions. In general, diesel engines burn fuel more efficiently than gasoline engines and emit lower

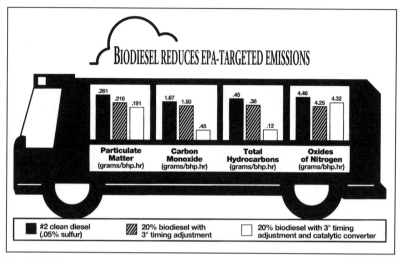

Fig. 4-1 Biodiesel Reduces EPA-Targeted Emissions. Source: 1993 National Soy Diesel Development Board; ORTECH International/Fosseen Manufacturing and Development

hydrocarbons, carbon monoxide, and nitrogen oxides. Particulates, however, are a major problem in compression-ignition engines. Even low-sulfur diesel fuel has much more sulfur, 40 times as much as gasoline, and heavier hydrocarbons, especially aromatics, which do not burn easily, resulting in particulate matter and visible smoke.

This smoke factor has helped biodiesel gain attention as both an additive and a substitute fuel because it offers significant emissions reductions in hydrocarbons and carbon monoxide with no modifications to bus engines and transit infrastructures. Another environmental plus for biodiesel and low-sulfur diesel is in the fuel economy area. Diesel engines have a much longer range than gasoline-powered ones.

Biodiesel, as a renewable biofuel, adds no net greenhouse gases to the atmosphere, if one excludes the fossil energy used to produce the feedstock and in the conversion process.[7]

Still, one large obstacle facing biodiesel is that the fuel tends to increase nitrogen oxides, an ozone precursor. Another problem is that biodiesel's relatively recent introduction has left a gray policy area for fleet operators and others wanting to use biodiesel to meet federal clean air requirements and/or the Energy Policy Act of 1990 (EPACT). Among the regulatory requirements biodiesel must meet is a Clean Air Act requirement that all fuels sold for use in compression-ignition engines be either "substantially similar" to diesel fuel or receive a waiver from the requirement. EPA has temporarily granted biodiesel permission to be sold legally, but the issue will need to be resolved before the fuel can be marketed on a nationwide basis. The Department of Energy, meanwhile, has not formally recognized biodiesel as an alternative fuel, although DOE officials have referred to biodiesel as an alternative. To meet EPACT guidelines, fleet managers must use an approved alternative fuel.

CONCLUSION

As a substitute for low-sulfur and traditional diesel fuel, biodiesel made from biomass is superior from an emissions standpoint but is inferior from a fuel cost standpoint. Complicating the picture is that the fuel still must meet various government agency requirements under EPA and DOE to qualify as an alternative fuel. Where it is likely to hold the most promise is in a blended diesel fuel. In this way, it can be used to reduce harmful emission and extend domestic fuel supplies, which could become an issue when crude oil prices escalate.

Perhaps the greatest advantage biodiesel has over other fuels is the overall savings a potential alternative fuels user can achieve, considering the investment required to modify or purchase expensive specialty vehicles such as transit buses and heavy-duty hauling equipment to meet pollution standards. While biodiesel may be expensive compared to conventional fuels, fleet managers may find it cost-effective when compared to the cost of additional pollution control devices. Biodiesel supporters call this a "least cost per attainment" solution.

Another plus for biodiesel is that there is no retraining required for motorists, mechanics, or refueling station employees. Both biodiesel and low-sulfur diesel are similar enough to conventional diesel that they can be assimilated into the existing petroleum fuel infrastructure so that it is completely transparent to consumers, or so is the hope of regulators and suppliers. This appears to be the situation today, although as with reformulated gasoline, the introduction of low-sulfur diesel suffered some growing pains, including price spikes 10 cents above traditional diesel and short supplies in some areas. But now that refiners have some experience under their belts, the following years have been uneventful, which is how motor fuel distribution should be. Whether biodiesel will successfully, and with little fanfare, carve a niche into the heavy-duty diesel engine market is still an open question.

References

1. Merrill, Peter Nathaniel. "Biodiesel: A Low Cost Option," *Utility Fleet Management*, September 1993.
2. Adler, Kevin. "Diesel Markets Undergo Change, Says World Conference Panel," *Octane Week*, April 11, 1994.
3. EIA. *Alternatives to Traditional Transportation Fuels: An Overview*, June 1994.
4. University of Wisconsin. *Alternative Fuels Refueling Station Infrastructure Study*, April 1, 1994.
5. Lorenzetti, Maureen. "How to Get Into the Business of Supplying Alternative Fuels," *Independent Gasoline Marketing Magazine*, July/August 1994.
6. Lorenzetti, Maureen. "EPA Not Expected to Require Reformulated Diesel," *Platt's Oilgram News*, Aug. 31 1995.
7. Holmberg, William. "Technology Advances & New Applications for Biodiesel," *1994 Conference on Clean Air Act Implementation and Reformulated Gasolines*, October 9-11, 1994.

CHAPTER FIVE

FUEL METHANOL

F uel-grade methanol is an alternative fuel that can be produced from fossil sources such as natural gas or from agricultural feedstocks, including biomass. Some associate methanol with the Indianapolis 500, where it has been used by professional auto racers for the last 30 years for its high octane level and performance advantages over gasoline. But the fuel has also been used over the years for its clean-burning characteristics or as a replacement for gasoline during crude oil supply shortages.

In the United States, it is most often used as a chemical feedstock for the oxygenate MTBE. But it also has been considered mainly by state governments and fleet operators in M-85 for use in flexible fuel vehicles (FFVs).

Photo 5-1   Flexible fuel vehicle

FFVs are cars that can operate on any combination of alcohol or gasoline. Still another highly promising avenue for methanol use is in powering heavy-duty vehicles such as buses and trucks, according to the American Methanol Institute. In the national effort to improve air quality in the smoggiest cities by reducing toxic emissions, methanol has been considered as a leading clean-burning alternative to other more common alternative fuels because it does not contribute to ground-level ozone, can be refueled at the retail level like conventional gasoline, and historically has been less expensive than gasoline. While most methanol fuel does not qualify for government subsidies like ethanol, factors such as supply and feedstock cost allow methanol to remain competitive as a chemical feedstock and as a neat fuel.

Another area of promise for expanded methanol fuel use

is in the development of fuel cell technology, which some see as a solution to the inadequacy of battery power in electric vehicles. Fuel cells make electric current by combining hydrogen and oxygen, with the hydrogen coming from methanol. This next generation of fuel cells has the potential to be used in a wide variety of vehicles, providing a driving range similar to conventional vehicles but using about half the energy (see Chapter 10).

However, as with any alternative fuel today, methanol has several disadvantages that are hampering its use as a replacement for gasoline. For example, methanol's energy density is about half that of gasoline, reducing the range a vehicle can travel on an equivalent tank of gasoline. Also, current methanol-powered vehicles that run on M100 may be difficult to start when outside temperatures dip below 45 degrees Fahrenheit because of methanol's lower vapor pressure and single boiling point. Price and availability are other issues that must be addressed. When reformulated gasolines were first introduced, demand for methanol as a petrochemical feedstock for MTBE skyrocketed, and supplies tightened severely enough to dramatically boost prices. The cost of a gallon of methanol nearly doubled in the space of one year. Additional capacity has since been added and prices have fallen, but some fuel providers question whether the methanol industry is ready to make the expensive capital investment necessary to ensure adequate neat fuel supplies if demand for M85 or M100 were to expand dramatically beyond current levels.

FUEL CHARACTERISTICS

Methyl alcohol (CH_3OH) is a colorless, poisonous liquid with essentially no odor and very little taste.[1] Chemically speaking, it is the simplest alcohol and boils at 64.7 Celsius. It mixes easily with water and most organic liquids, including gasoline, and is extremely flammable. More significantly, it burns with a nearly invisible flame. Typically, methanol is produced commercially by the catalyzed reaction of hydrogen and carbon monoxide, using natural gas as a feedstock. At one time, it was produced by distilling wood, hence the name "wood alcohol."

Nearly all methanol used in the transportation sector alone or in MTBE is derived from natural gas. Natural gas is inexpensive and convenient to use for methanol production, which is often set up near other petrochemical facilities. Natural gas resources in the United States are estimated to be from 300 to 500 trillion cubic feet (Tcf). By comparison, current U.S. consumption of natural gas is about 20 Tcf/year. There are obvious advantages to using natural gas for methanol production. Namely, the technology for making the alcohol from gas is well-established and requires only efficiency improvements. The only significant disadvantage to relying on natural gas to make methanol is political, not scientific.

Current federal law provides alcohol tax exemptions to both ethanol and methanol, provided the feedstock is agriculturally based. Biomass resources, whether they be corn or landfill waste, are potentially unlimited, but production techniques are still not cost-competitive with natural gas. Another little-used feedstock for methanol is coal. Coal reserves in the United States are vast — about four trillion tons — but again the cost of production is significantly higher than natural gas.

Methanol has a higher octane rating than gasoline, making the alcohol well suited as a spark ignition fuel.[2] Since it has a relatively high octane value, it has a correspondingly low cetane level of around 15, versus 40 to 55 for diesel. Methanol alone does not provide adequate lubrication to high-pressure pumps used to inject fuel into the compressed air in the combustion chamber of a compression engine, but this problem can be overcome through the use of fuel additives or the addition of an ignition source such as a glow plug in some methanol engines.

In order to avoid these engine performance problems, methanol fuels usually contain 15 percent gasoline. Adding gasoline to methanol also improves flame visibility and increases the energy density of the fuel.

STORAGE AND DISTRIBUTION

Because methanol is a liquid fuel just like gasoline, in theory there should be no major technical or institutional barriers to impede the expansion of a methanol distribution network. But in practice, the oil industry is resistant to modifying its infrastructure to accommodate any alcohol fuel even though both ethanol and methanol are liquids at ambient temperatures and atmospheric pressures. There are several reasons for this. At the pipeline level, operators fear alcohol fuels will suffer water contamination, phase separation in the cases of M85 or E85, or cross-contamination from other petroleum products moving along the system. Some pipeline and terminal operators also fear that the corrosive nature of methanol fuels will require expensive new investment in storage equipment.

As a result, there is very little infrastructure in place to serve neat methanol vehicles. One of the largest systems exists in California, where the state has embarked on public-private partnerships with the oil industry to install about 50 methanol fueling sites. Typically, methanol is dispensed at gasoline stations in methanol-compatible gasoline pumps. To prevent drivers of gasoline vehicles from mistakenly fueling with methanol, the state uses a card-key system to limit access to methanol dispensers at public-access fueling stations. Some methanol vehicle operators perceive this system as a hindrance, so the industry is searching for a more user-friendly system to limit access. Since flex-fuel methanol vehicles also use gasoline, using a dedicated fueling nozzle is not considered feasible. Instead, efforts are focused on using a standard fueling nozzle which incorporates an interlock system that would permit only methanol-compatible vehicles to fuel with methanol (a similar system could be used for ethanol).

Upstream from the service station, unit costs for transporting methanol are higher than those for transporting petroleum fuels, California state officials have found. Part of these costs are incurred in preparing blended M85. At present, fuel methanol haulers typically have to make at least two stops, one to load methanol at a chemical terminal facility and one at a gasoline terminal to add blending gasoline. The ability to load both fuel methanol and gasoline at the same loading rack would reduce the wholesale cost of M85.

Additional terminals and storage capacity, open on a round-the-clock basis, are needed to ensure a reliable supply of M85 and M100 and to make the fuel more competitive with gasoline. Some critical applications for this fuel, such as mass transit, are presently operating with slim margins of supply. At present, fuel methanol is hauled from terminal to dispenser storage, using tank trucks and trailers that otherwise carry conventional gasoline and diesel fuels. Under current operating practices, the tank trailers must be free of pools or puddles of hydrocarbon fuels before loading with methanol. This procedure is necessary to ensure that the methanol fuel is well within the specifications for current methanol and fuel-flexible engines. A dedicated fleet for transport of M85/M100 fuels would eliminate the opportunity for product contamination. Such a dedicated fleet would also increase the likelihood of having knowledgeable fuel handlers to maintain product quality and prevent bulk misfueling. Tanker truck trailers, however, must operate round-the-clock to be economically viable, but demand does not support such an operation. About 1,400 truckloads, or less than 30 loads a week, of fuel methanol are delivered each year for the entire state of California.

BUILDING CODES

In California, fire codes treat methanol in the same way as gasoline, and local regulations already in place require that new underground tanks at service stations be methanol-compatible. In general, the regulations governing conventional fuel storage are also applicable to ethanol refueling stations. Some additional requirements or delays may occur if local officials are not familiar with the fuel characteristics. One good way to avoid problems is to provide local officials with sufficient information early in the retail planning process.[4] On the retail level, station operators should be aware of the need for special equipment which applies primarily to methanol and methanol fuel blends rather than to ethanol fuels, due to the aggressive nature of methanol as a solvent and to its corrosion potential. The flammability limits for methanol and methanol/gasoline blends differ from those for gasoline alone, creating the need for appropriate equipment design features to eliminate fire and explosion hazards. For example, because M100 has explosive vapor limits at normal ambient temperatures within closed vessels, e.g., fuel tanks, measures are required to prevent ignition within the vessel and flame-front propagation from ignition sources outside through vessel openings. Standard specifications should be established for flame arrestors and other appropriate measures on fuel storage tanks, vents, and pumps. M85 does not have this drawback, due to the vapor pressure of the gasoline fraction, which results in a saturated vapor inside closed vessels under normal ambient temperature conditions.

The methanol industry and equipment manufacturers are the most likely parties for obtaining revisions to engineering and materials standards. The Canadian Oxygenated Fuels Association and retailers such as Mohawk Norwood, a fuel retailer, and Ford Motor Company have established their own equipment specifications for methanol fueling stations.

According to the National Fire Protection Association (NFPA), the electrical specifications for ethanol and methanol are the same as those for gasoline. This issue is covered in American National Standards Institute (ANSI)/NFPA 30, *Flammable and Combustible Liquids Code,* and ANSI/NFPA 30A, *Automotive and Marine Service Station Code.* Gasoline, ethanol, and methanol fueling facilities are required to have explosion proof electrical dispensing equipment. Unfortunately, the potential need for electrical specifications for methanol fueling systems cannot be fully resolved until more is understood about the role of methanol conductivity.

MATERIAL COMPATIBILITY ISSUES

The requirements for E85 or M85 underground storage and dispensing systems are similar to conventional petroleum service station installations, with some material substitutions as the main difference. Both alcohols are corrosive, although methanol is more so, and the use of improper materials could result in fuel contamination, equipment failure in the dispensing and metering equipment, or leaks in the alcohol storage and distribution system. For this reason, public-access refueling sites often require fleet cards so misfueling does not occur. Also, there are slight differences in material compatibility which preclude marketers from randomly switching a pump from M85 to E85.

Ethanol is noncorrosive to most metals at ambient temperature except for the following compounds: terne, steel, magnesium, and aluminum. Sensitive nonmetals include natural rubber, polyurethane, cork gasket material, polyester-bonded fiberglass laminate, polyvinyl chloride, and certain other plastics, including polyanides and methylmethacrylate.

By comparison, anhydrous methanol is noncorrosive to most metals at ambient temperatures except lead, magnesium, and platinum. However, coatings of copper or copper alloys, zinc, or aluminum should also be avoided because coatings of these materials are attacked, though more slowly. Mild steel alloys have been used successfully in many alcohol handling systems and are therefore recommended for tanks, pipes, and other system components. Nonmetallic components like resins, nylons, and rubbers such as nitrile, polyurethane, polypropylene, and neoprene elastomers have also been used for ethanol equipment without problems.

Components to outfit an E85 or M85 station are similar to a gasoline station, with the following caveats: The underground storage tank should be double walled with interstitial space and automatic leak detection, and cathodic protection is recommended for steel tanks placed underground. Also, marketers should check the age and composition of an existing tank before adding alcohols. A new fiberglass is perfectly acceptable to use, but tanks installed before 1987 are not compatible with these fuels and will need to be replaced. Product piping should have ethanol or methanol compatible filters and pump seals, and trim must be compatible with the alcohol of choice. The dispenser and metering equipment, including flow regulator, spin-on dispenser fuel filter, dispenser hose, and dispensing nozzle, all must be modified for ethanol or methanol. Many of the nozzles now sold on the market are nickel plated and typically cost about twice the amount of the gasoline version.

SAFETY AND ENVIRONMENTAL ISSUES

From an environmental standpoint, a methanol spill is much less damaging than a gasoline leak because alcohols are highly water soluble and disperse rapidly. Unconfined alcohol can contaminate ground water and cannot be easily extracted from the water. But compared to gasoline, methanol can be degraded by naturally occurring bacteria fairly rapidly. According to the EPA, given aerobic conditions, methanol can be degraded in a matter of weeks, rather than years for gasoline.

FUEL TOXICITY

Methanol is toxic via all modes of contact, including inhalation, ingestions, and skin penetration, and poses challenging handling considerations, given that the toxic effects of methanol can be seen by ingestion of as little as three to four teaspoons. Fatal doses typically require a victim swallowing 20 to 40 teaspoons, however. Methanol ingestion symptoms vary depending on the individual, but 10 to 48 hours after the alcohol has been swallowed, a number of systems can occur, including headache, nausea, blurred vision, or leg and back pain. If exposure is severe enough, symptoms can progress to temporary or permanent blindness, coma, and possible death from respiratory failure.

While the toxic effects of the fuel cannot be discounted, it's also important to note that accidental ingestion of methanol is extremely rare, given that suppliers include additives that allow for taste and odor detection.

FLAMMABILITY ISSUES

The fire and explosion risk of any fuel depends on a number of factors, including the volatility and ignition temperature of a fuel and the amount of oxygen present. When a methanol fire does occur, the energy content of the alcohol and the heat required to vaporize the fuel means the fire burns less intensely and may cause less damage than a gasoline fire.

Methanol has a vapor pressure of 7.3 psi, slightly lower than gasoline, which is 8 to 15 psi, depending on season and regulatory adjustments, so methanol is less likely to ignite than gasoline in open spaces. However, in closed spaces such as fuel tanks, M100 vapors form a combustible mixture because the vapors form a wide range of mixtures of fuel and air that could be ignited by a spark. Fortunately, the required addition of denaturants, such as gasoline, takes care of this problem because it raises the fuel's vapor pressure beyond the flammability limit, so the fuel mixture in storage tanks and pumps is too rich to ignite.

Materials typically used to extinguish petroleum-based fires are also good tools for a methanol fire. Because water easily blends with methanol, water will spread, not extinguish, a methanol fire. Instead, extinguishers such as dry chemicals, carbon dioxide, or alcohol-resistant foams should be used.

If methanol fuel use expanded to more popular levels outside of California, fire-fighting tactics would have to accommodate certain methanol characteristics. In its pure state, methanol burns with a colorless flame, so a methanol fire in open daylight is hard to see. In addition, it is somewhat volatile, with a Reid vapor pressure of 4.5 for M100, compared to 7.5 or higher for gasoline, and it has a relatively low flash point. M85 has a higher volatility than M100, due to the gasoline fraction. Currently, there is no NFPA code for methanol although there are codes pending. Fire fighters would have to use an alcohol-resistant fire fighting foam in a methanol emergency. Fire-fighting tactics, however, are not a code issue. Nevertheless, the flammability issues suggest regulatory codes need to cover location of methanol and ethanol storage tanks at a fueling site. NFPA 30 and 30A cover this issue.[5]

PUBLIC HEALTH

As the use of alcohol-fueled vehicles grows, various local, state, and federal governments provide information to educate the public regarding the particular features of the fuel. In particular, consumers need to be aware of toxicity and flammability issues and the importance of safety features to avoid contact, inhalation, ingestion, and vehicle safety hazards. Regulators are expanding these public education efforts to ensure that M85, M100, and denatured ethanol are not used inappropriately for household applications, such as solvents and small-engine fuels. These concerns can be readily addressed through education and labeling, and do not represent a major barrier to infrastructure development.

ECONOMICS

SUPPLY

Perhaps the single largest problem hampering the expansion of methanol as an alternative fuel is price. Because methanol can be made from numerous feedstocks, including natural gas, coal, and theoretically, biomass, methanol feedstock supplies are extensive. The current total North American supply of methanol is 2.6 billion gallons, and an 800-million-gallon increase in supply from sources in North and South America also will be online soon. Other large capacity increases are expected from the Middle East because of the availability of inexpensive natural gas.

World methanol supplies are expected to grow to 33 million metric tons by the year 2000, according to industry estimates published by methanol expert James Crocco, with total demand projected to reach 31.2 million metric tons. Of that demand, close to one-third will be used to make MTBE, about one-half will supply the chemical market for formaldehyde and acetic acid, with the remaining amount expected to go to specialty chemicals and neat fuels.

Given world capacity expectations, international suppliers could easily support an expanded fuel methanol refueling network that is double or even triple current demand levels. But methanol producers themselves don't appear willing to commit to capacity levels beyond what is expected to keep pace with demand for normal and traditional chemicals, solvents, and MTBE uses, with enough overcapacity to provide a safe cushion.[6]

For now, demand for methanol fuels could easily be met within U.S. borders. In California, for example, flexible fuel vehicles operating in the state use the equivalent of less than 1 percent of the U.S. methanol supply. The rest is consumed for other uses, mainly as a feedstock for the oxygenate MTBE, which is used in reformulated gasolines. In fact, U.S. and worldwide investments in new large-scale methanol plants were stalled until the mid-1990s since the industry was fearful of repeating the mistakes it made during the energy crises of the 1970s.

In the late 1970s and early 1980s, a number of new methanol plants were conceived and built throughout the world, hoping to take advantage of the expectation that U.S. consumers would want to use methanol fuels because crude oil was expected to reach $60/bbl. But the U.S. industry overbuilt, constructing 9 million tons of capacity when demand only reached 1 million tons because crude forecast proved woefully incorrect.

Today, lingering questions by the public on the safety of MTBE could dampen the rosy forecasts for the ether in the next three to five years and could spur methanol producers to encourage 100 percent methanol and methanol-gasoline blends that were used in the 1970s during the energy crises. Moving forward on that path will require methanol supporters to provide the public and fuel distributors with correct information to avoid confusion and bad publicity. For example, in the mid-1980s a series of limited incidences of poor engine performance occurred due to improper fuel handling, and a lack of public information on the problem has led motorists to continue to mistrust alcohol fuels even though producers have more safeguards to protect fuel quality. Additionally, blending methanol directly into gasoline does not make economic sense from a producer standpoint unless crude oil is at least $25 to $30 per barrel, with plant profit based solely on the cash cost of methanol production, which for a 3.5 million tons/year U.S. Gulf Coast plant can typically be anywhere from 25 to 32 cents per gallon, depending on feedstock cost and plant location.

PRICING ISSUES

Because the market demand for methanol vehicles will be driven by fuel price, the implication of higher-than-expected methanol prices may therefore further constrain market growth for methanol-powered vehicles. One way this issue may be addressed is through importing overseas production, which often is sourced from the politically volatile Middle East. As a result, it's unlikely U.S. policymakers will want to be put in a position of expanding a replacement to gasoline that could pose additional energy security risks, even if the fuel in question is environmentally superior to petroleum.

Another more realistic long-term option to expand neat methanol fuels is to use biomass, a virtually inexhaustible feedstock if properly managed, once production costs for this source dropped to competitive levels. Note that fuel cost and demand are in part a function of the existing tax structure and are subject to change.

Some promising work is being done by The National Renewable Energy Laboratory (NREL), which is developing technologies to produce both ethanol and methanol from biomass, including using landfill waste. NREL estimates that 2.45 million tons of biomass annually are available for conversion to alcohol fuels in the United States. This volume of biomass could theoretically produce approximately 270 billion gallons of ethanol or 455 billion gallons of methanol. Converting biomass to alcohol is more problematic than converting corn or sugar cane because biomass is largely composed of cellulose rather than simple sugars. But NREL has developed a process called *simultaneous saccharification and fermentation* that optimizes the conversion of cellulose. Under a cooperative research and development agreement with Amoco, NREL has built a pilot plant that converts one ton/day of municipal solid waste into ethanol. Methanol can also be produced using the process. NREL also is developing a thermochemical gasification process to make methanol from bagasse (sugar cane fibers). Government officials say that the current estimated cost of producing methanol from biomass through gasification is 84 cents per gallon. The goal of the NREL program is to reduce this cost to 50 cents per gallon, but the lab has not yet developed information on the cost-effectiveness and environmental impacts of producing methanol from biomass relative to other methods of methanol production

DEMAND

In the eyes of methanol supporters, such as the California Energy Commission, while methanol fuels cannot compete directly with gasoline on a gallon-for-gallon energy equivalent basis, other public policy benefits, particularly cleaner air, pay for their additional cost. But beyond state and federal government approval of methanol fuels, the industry has suffered because there is no major product champion in either the automotive or fuels and energy industries that is advocating methanol. Therefore, without a major push from private industry, new investments in methanol production facilities targeted at the transportation fuels market will be minimal.

Today, demand for methanol reflects this initial prognosis. MTBE and other chemical derivatives consume the lion's share of methanol made in North America, while methanol used in M85 and M100 represents less than 0.2 percent of demand. Vehicles in California consumed most of the M85 produced in North America, and M100 use has been negligible. According to CEC, in 1993, 12 million gallons of methanol were used directly as a transportation fuel in California, while the total statewide use of methanol for all fuel and fuel additives, including MTBE, was estimated at 212 million gallons. The majority of the fuel-grade methanol used in the state comes from sources in Canada, Colorado, and the U.S. Gulf Coast. Still, California has already experienced supply problems to meet demand by its demonstration fleet at prices that are competitive with gasoline at the pump, due in part to the higher-priced MTBE feedstock market. In 1994, about 10,000 light-duty FFVs were operating in the state, but surveys done by state fleet operators showed that these FFVs were predominantly operated on gasoline. Fleets are the dominant FFV users, although the use of FFVs by private individuals has been increasing, mainly in California. Additionally, some 500

heavy-duty methanol vehicles are operating in the state, primarily in transit or school bus applications. The Los
Angeles County Metropolitan Transit Authority currently
operates 333 methanol transit buses on M100. In addition,
16 school districts in the state operate a total of 150
methanol-powered school buses, provided as part of the
Katz Clean Safe School Bus Program. Most of these buses
operate on M85, but some are being converted to M100.

ENVIRONMENT

From a mobile source standpoint, methanol's high
heat of vaporization results in lower peak flame temperatures than gasoline, and lower tailpipe emissions
of ground-level ozone precursors, carbon monoxide and
hydrocarbons. Emissions of primary concern from alcohol
vehicles are aldehydes, and in particular, formaldehyde.
Methanol vehicles emit greater amounts of the chemical than
ethanol vehicles. Conversely, ethanol vehicles are higher
emitters than conventional vehicles. While formaldehyde is a
probable carcinogen, in the eyes of EPA the greater health
risk is from ground level ozone and carbon monoxide, which
scars lung tissue and can increase the risk of some respiratory
cancers. Methanol and M85 offer some advantages over gasoline in these areas, but they are not as clean burning as other
available alternative fuels, namely electric and hydrogen powered vehicles that are technologically immature but offer the
benefit of having zero emissions.

GLOBAL WARMING

What role replacement fuels may play in reducing greenhouse gases is an area of extreme controversy. Scientists generally agree that methanol and other alternative fuels, such has compressed natural gas, ethanol, propane, and hydrogen, emit less carbon dioxide than gasoline. But there is still a lively political and scientific debate on what fuels provide superior protections against the threat of global climate change. Both CNG and methanol fuels, for example, generate considerably less carbon dioxide than the gasoline fuel cycle but emit considerably more methane, another greenhouse gas. Uncertainty over the warming potential of methane relative to carbon dioxide further complicates meaningful debate.

CONCLUSION

Of all the alternative fuels, methanol is the least expensive and the most similar to gasoline in terms of shipping requirements and vehicle considerations. But the increased use of methanol as a motor fuel and the growth in the number of FFVs on the highways of the future will be linked to increasing the supply of methanol and lowering its price to a level that is competitive with other fuels. For methanol fuels to expand, investors must perceive a stable, reliably increasing demand to justify new plant construction.

Both methanol fuel suppliers and their customers

concede that developing methanol as a neat fuel will be a slow process and may never reach appreciable marketshare compared to gasoline unless crude values suddenly sky-rocketed to $30/bbl and stayed there indefinitly. But that is not to say that methanol can't provide a niche role for some fleet managers. While fuel methanol will cost more than traditional fuels, in some cases the clean air benefits and limited refueling adjustments may make the alcohol competitive with other replacement fuels.

References

1. EIA. *Alternatives to Traditional Transportation Fuels: An Overview*, June 1994.
2. University of Wisconsin. *Alternative Fuels Refueling Station Infrastructure Study*, April 1, 1994.
3. Calfuels Plan. *Developing an Infrastructure Plan for Alternative Fuel Vehicles*, California Energy Commission, September 1994.
4. Science Applications International Corporation. *Introduction to Alternative Fuel Vehicles*, January 1993.
5. Lorenzetti, Maureen. "How to Get Into the Business of Supplying Alternative Fuels," *Independent Gasoline Marketing Magazine*, July/August 1994.
6. Crocco, James. *Availability and Demand for Methanol for Use as a Neat Fuel and for Reformulated Gasoline*, World Conference on Clean Fuels and Air Quality Control, October 6-8, 199?, Washington, D.C.

C H A P T E R   S I X

ETHANOL

Fuel grade ethanol, made almost exclusively from corn in the United States, enjoys popularity among Midwest motorists because it is seen as a way to boost farm income. But for independent refiners and gasoline marketers, the alcohol has been popular less for its touted agricultural benefits than because it acts as a reasonably-priced gasoline extender and low-cost octane enhancer, thanks to generous federal and state subsidies. Still, even with attractive tax incentives, ethanol fuels outside the Midwest have had a hard time competing on a gallon-for-gallon basis with traditional motor fuels such as gasoline and diesel because of high transportation costs associated with its use and cost of production. Ethanol is most commonly used as a fuel additive for conventional or reformulated gasoline, as discussed in Chapter Three. Adding fuel ethanol in these amounts requires no modifications to the vehicle or refueling system. Ethanol also can be used as a replacement for gasoline, but supply and price have limited high volume ethanol fuel to very small selected markets where price is not a factor. Ethanol has also suffered from consumer acceptance problems, which came about from isolated fuel quality problems brought on by inexperienced fuel suppliers that improperly mixed alcohol/gasoline levels.

In today's marketplace, however, procedures such as in-line computer blending at the terminal level have effectively eliminated this problem. For the purposes of this chapter, the discussion will be limited to the use of ethanol when it is used as a gasoline additive, sometimes called *gasohol* or E10, or as the replacement fuel E85, which is 85 percent denatured ethanol and 15 percent unleaded gasoline with additives.

FUEL CHARACTERISTICS

Ethanol (CH_3CH_2OH), otherwise known as ethyl alcohol, is a clear, colorless, flammable, oxygenated hydrocarbon, with a boiling point of 78.5 degrees Celsius. However, in the presence of water, it forms a binary azeotrope, with a boiling point of 78.15 degrees Celsius at a composition of 95.57 percent by weight ethanol. Ethanol is much more volatile than gasoline, so it does not vaporize as readily as gasoline in cold weather temperatures, which can lead to driveability concerns such as stalling. For this reason, the cold start limitation is overcome in current alternative fuel vehicle designs by adding 15 percent gasoline by volume to the ethanol.[1]

Ethanol, in its most familiar form, is produced by the fermentation of biomass, often agricultural products like grapes, corn, or barley, which are all feedstocks for wine, beer, whiskey, and other spirits. To make fuel grade ethanol, the fermented product, which typically contains 10 percent ethanol and 90 percent water, is distilled to separate the alcohol from the water, just as a refiner separates crude oil into different boiling range fractions to obtain different products. Unlike beverage ethanol production, an ethanol producer will then dehydrate the remaining water out of the alcohol using technology such as a molecular sieve, or he will add another distillation step to ensure no water is present in the fuel that could cause an engine to stall or even stop. Ethanol manufacturers are also required by federal law to add a denaturant such as gasoline to the fuel-grade alcohol so there is no incentive to drink one's fuel instead of driving it. But even without the denaturant, fuel-grade alcohol almost always contains toxic impurities like higher alcohols and fuel oils that a whiskey distiller or a perfumer would have to refine out of his product.

In fact, most industrial uses which require a high-level

of purity, like pharmaceuticals and cosmetics, are produced by hydrating the petrochemical ethylene. This occurs by mixing ethylene with steam at 60 to 70 atmospheres and about 300 degrees Celsius over a phosphoric-acid catalyst supported on diatomaceous earth as seen in the reaction below[2]:

$$CH_2 = CH_2 + H_2O \quad CH_3CH_2OH$$

Ethylene   Water   Ethanol

But for the purposes of this discussion, the focus will be on fuel ethanol made from corn, as shown in the reaction below:

$$C_6H_{10}O_5 + H_2O \quad 2C_2H_5OH + 2CO_2$$

Starch   Water   Ethanol   Carbon Dioxide

The reason for this seemingly narrow focus is that current tax incentives which encourage fuel ethanol only apply to fuel produced from renewable energy sources like agricultural or biomass. Currently, more than 90 percent of the fuel ethanol produced in the United States is made from corn, although feedstocks such as sugar cane, wheat, and even artichokes have been used. Corn continues to dominate because it is already used extensively for food and feed and enjoys a price support system funded by the federal government.

Growing interest in ethanol produced from wood and

municipal waste has encouraged numerous research both by the federal government and private industry, but neither feedstock can currently compete on an economic basis with corn.

STORAGE AND DISTRIBUTION

Regardless of supply origin, ethanol still is confined by an existing U.S. petroleum product infrastructure not designed for alcohol distribution. Underground storage tanks and underground pipelines create natural condensation and attract continuous amounts of water. For this reason, ethanol's affinity for water and the resulting potential phase separation have led refiners to resist retrofitting pipelines and tanks to meet alcohol's special needs.

In most cases, refiners have labeled these infrastructure problems as burdensome or expensive, but not unsurmountable. Except for water tolerance, ethanol handling is similar to other motor fuels. In the future, shipper encouragement and the development of an emulsifier agent could solve this problem. Handling ethanol blends is not a serious technical problem which cannot be overcome, but rather the preference and lack of refiner demand for ethanol which gives pipeline companies and refineries reason not to incur extra costs or liability of handling ethanol blends. But in the future, it is likely clean air laws and ethanol blending economics will cause some pipeline companies to reconsider ethanol and ethanol blend shipments. Some pipeline companies have already made additional investments such as changing seals and other pipeline parts to allow for the transport of neat MTBE because of market demands for that product.

Further demand for the ethanol ether ETBE might finally allow ethanol to be included in the pipeline system since refiners may find it cheaper to retrofit existing pipelines to ship the alcohol to their plant gate instead of relying on the current rail, truck, and barge system currently used to transport ethanol from the Midwest to outside locations.

Today, ethanol used for gasoline blending can be purchased at the ethanol facility or at a gasoline wholesale terminal at the end of a pipeline. At a terminal, proprietary brand additives, detergent packages, other oxygenates, and various gasoline components are blended to make various grades of gasoline. As is the case with gasoline, ethanol is delivered to retail gasoline outlets via 8,000 gallon dealer tank wagons (DTW). Other than proper maintenance and standard housekeeping procedures employed at the wholesale and retail levels standard with any motor fuel, ethanol poses no technical impediments after leaving the wholesale terminal.

Many wholesale gasoline terminals no longer have rail facilities as a result of increased pipeline and barge shipments over the past several decades. Due to the lack of rail facilities at the wholesale level, higher volumes of ethanol place added pressure on the scheduling of ethanol shipments via DTWs in areas without the ability to accept larger loads of ethanol by rail or barge. The use of ethanol in the manufacture of ETBE and its inclusion in gasoline is expected to assist in eliminating the need for separate handling and storage now associated with ethanol blends.

BUILDING CODES

In general, the regulations governing conventional fuel storage are also applicable to ethanol refueling stations. Some additional requirements or delays may occur if local officials are not familiar with the fuel characteristics. One good way to avoid problems is to provide local officials with sufficient information early in the retail planning process.

MATERIAL COMPATIBILITY ISSUES

The requirements for E85 or M85 underground storage and dispensing systems are similar to conventional petroleum service station installations, with some material substitutions as the main difference. Both alcohols are corrosive, although methanol is more so, and the use of improper materials could result in fuel contamination, equipment failure in the dispensing and metering equipment, or leaks in the alcohol storage and distribution system. For this reason, public-access refueling sites often require fleet cards so misfueling does not occur. Also, there are slight differences in material compatibility which preclude marketers from randomly switching a pump from M85 to E85.

Ethanol is noncorrosive to most metals at ambient temperature except for the following compounds: terne, steel, magnesium, and aluminum. Sensitive nonmetals include natural rubber, polyurethane, cork gasket material, polyester-bonded fiberglass laminate, polyvinyl chloride, and certain other plastics, including polyanides and methylmethacrylate.

By comparison, anhydrous methanol is noncorrosive to most metals at ambient temperatures except lead, magnesium and platinum. However, coatings of copper or copper alloys, zinc, or aluminum should also be avoided because coatings of these materials are attacked, though more slowly. Mild steel alloys have been used successfully in many alcohol handling systems and are therefore recommended for tanks, pipes, and other system components. Nonmetallic components like resins, nylons, and rubbers such as nitrile, polyurethane, polypropylene, and neoprene elastomers have also been used for ethanol equipment without problems.

Components to outfit an E85 or M85 station are similar to a gasoline station, except the underground storage tank should be double walled with interstitial space and automatic leak detection, and cathodic protection is recommended for steel tanks placed underground. Also, marketers should check the age and composition of an existing tank before adding alcohols. A new fiberglass is perfectly acceptable to use, but tanks installed before 1987 are not compatible with these fuels and will need to be replaced. Product piping should have ethanol or methanol compatible filters and pump seals, and trim must be compatible with the alcohol of choice. The dispenser and metering equipment, including flow regulator, spin-on dispenser fuel filter, dispenser hose, and dispensing nozzle, all must be modified for ethanol or methanol. Many of the nozzles now sold on the market are nickel plated and typically cost about twice the amount of the gasoline version.

SAFETY AND ENVIRONMENTAL ISSUES

The overall spill of either methanol and ethanol is much less damaging than a gasoline leak because alcohols are highly water soluble and disperse rapidly. Ethanol fuels are toxic to marine, animal, and even human life in high concentrations, but they are much less toxic than other liquid fuels, including gasoline, diesel, and methanol. And since ethanol is less toxic to humans than gasoline in equal concentrations, a large ethanol spill has less of a chance of contaminating local water supplies. Unlike methanol, ethanol is not a dangerous poison if ingested and is in fact relatively nontoxic at low doses, and the presence of a denaturant renders it undrinkable in its fuel form. Since pure ethanol or high-volume ethanol blends available at refueling stations are premixed with gasoline, deliberate ingestion is less of a problem than accidental ingestion from siphoning accidents.

FLAMMABILITY ISSUES

The fire and explosion risk of any fuel depends on a number of factors, including the volatility and ignition temperature of a fuel and the amount of oxygen present. When an ethanol fire does occur, the energy content of the alcohol and the heat required to vaporize the fuel means the fire burns less intensely and may cause less damage than a gasoline fire.

Ethanol has a vapor pressure of 4.3 psi, lower than gasoline, which is 8 to 15 psi, depending on season and regulatory adjustments. This makes ethanol less likely to ignite than gasoline in open spaces. However, in closed spaces such as fuel tanks, E100 vapors form a combustible mixture because the vapors form a wide range of mixtures of fuel and air that could be ignited by a spark. Fortunately, the required addition of denaturants, such as gasoline, takes care of this problem because it raises the fuel's vapor pressure beyond the flammability limit so the fuel mixture in storage tanks and pumps is too rich to ignite.

Materials typically used to extinguish petroleum-based fires are also good tools for an ethanol fire. Because water easily blends with ethanol, water will spread, not extinguish, an ethanol fire. Instead, extinguishers such as dry chemicals, carbon dioxide, or alcohol-resistant foams should be used.

ECONOMICS

In the foreseeable future, fuel ethanol cannot compete with gasoline on a gallon-for-gallon basis. Ethanol economics become much more attractive when they are compared to other gasoline blendstocks on an octane or oxygenate basis, like toluene or MTBE. With the fuel ethanol industry in its infancy, producers are still at a cost disadvantage because of lower crude oil feedstock costs, economies of scale of gasoline refiners, and an immature infrastructure. Without federal incentives, ethanol could only compete in the U.S. motor fuel market under unique circumstances, such as when crude oil exceeds $30 per barrel or if new ethanol production technologies reduce costs below that of gasoline.

The ethanol industry currently receives a fuel tax exemption of 54 cents per gallon of ethanol. This means gasoline that contains 10 percent ethanol, for example, receives a 5.4 cent per gallon tax exemption from the federal excise tax. In addition, many states, primarily in the Midwest, provide ethanol producers with direct payments varying from 20 to 40 cents for each gallon of fuel grade ethanol they produce locally. Fuel suppliers that choose to market ethanol-blended gasoline can buy gasoline at a reduced rate in one of three ways: by registering as an ethanol blender, applying for an income tax credit based on the amount of ethanol blended into gasoline, or applying for a refund based on gasoline purchased at the full rate and later blended into ethanol.

Both oil and agricultural interests agree that ethanol has served a useful purpose over the years by helping to extend gasoline supplies during the energy crises of the 1970s and by providing a reasonably priced alternative to lead in the 1980s. It also serves as a high oxygenate seasoning for refiners' reformulated gasoline recipes. But at the same time, since ethanol fuels are a product of the agricultural, not petroleum sector, the ongoing efforts by the farm community to supplant petroleum with an agriculturally-derived fuel have led to many skirmishes between the two groups.

Interestingly, the U.S. government subsidizes various energy sources, including petroleum fuels. According to government figures, oil receives about $1 billion in federal energy subsidies, mainly aimed at increasing domestic production and exploration. In 1993 dollars, the total federal energy subsidy was $8 billion, based on a national energy bill of $475 billion.

However, oil receives a disproportionately small share of these energy subsidies — about 12 percent of the total — while providing close to 40 percent of the U.S. total energy supply. Further downstream, gasoline is taxed to help pay for other government programs, including a special highway trust fund used to maintain highways and bridges. For this reason, the petroleum industry argues that the $30 billion collected in a typical fiscal year in excise taxes more than offsets the oil industry's upstream subsidies.

This tax structure helps explain why oil companies often insist ethanol is the most heavily subsidized energy source in America.[3] Based on an annual consumption rate of 1.6 billion gallons, the total estimated federal ethanol subsidy is about $900 million.

Ethanol interests argue that their subsidy is cost-effective because it helps encourage a fuel that provides many indirect public policy benefits to the economy, such as cleaner air, better energy security, and higher farm income. And in the eyes of the ethanol industry, the government actually saves money with the subsidy. The combination of reduced farm program costs and increased income tax revenues from farmers results in a net gain to the U.S Treasury of $1.30 per gallon, which is more than double the government's initial investment.[4]

It's a debate likely to continue as long as there are gasoline-powered automobiles, which for all practical purposes, will mean decades. One point both parties agree on is that ethanol, whether it's made from corn or biomass, will need to be subsided for some years to come to remain a viable part of the motor fuel business.

SUPPLY

The U.S. fuel ethanol industry produces, mainly from corn, about one billion gallons per year of anhydrous product for downstream blending. Currently, all U.S. plants of 50 million gallons per year or more use a corn feedstock because of availability, transportation, and storage issues. The extremely large, established market for corn also allows established agriprocessors like Archer Daniels Midland and Cargill to maintain consistent production costs by hedging on commodity markets even though grain prices rise and fall with the unpredictability of summer thunderstorms.[5] And, just as crude oil run through a refinery can be made into any number of petroleum products, including gasoline, diesel fuel, asphalt, and fuel oil, when corn is converted into ethanol, other products are also produced. These include carbon dioxide, protein animal feeds, and sometimes corn oil and sweetener.

With the right combination of plant size, location, process, and investment, ethanol can compete with gasoline, provided that the tax credit is used. Also, if an ethanol plant is constructed immediately next to another facility, the cost per gallon can be lowered further because plant managers can take advantage of existing infrastructure and plant scale. One of the biggest factors which impacts ethanol cost is the size of the plant.

In the United States, two different processes — dry milling and wet milling — are used to produce ethanol from the starch in corn. Dry-milling is a method nearly identical to one used for whiskey. Indeed, many fuel alcohol plants now in production started off making whiskey or bourbon before fuel ethanol demand exploded in the early 1990s. Grain is hammermilled and cooked to make ethanol, with carbon dioxide and distiller's dried grains, an animal feed, produced as by products.

However, more than two-thirds of the major plants, with 66 percent of the U.S. ethanol capacity, use the wet milling process, which first became established by starch and corn sweetener manufacturers. A wet miller that adjusts his starch stream to make ethanol typically produces about 2.5 gallons of alcohol, 13 pounds of the animal feed corn gluten feed, 3.4 pounds of the animal feed corn gluten meal, and 1.7 pounds of corn oil from one bushel of corn. By comparison, dry milling one bushel of corn in an ethanol plant will produce about 2.5 gallons of fuel and 17 pounds of distiller's dried grains.

Wet milling requires a larger capital investment but generates more value-added products that reduce the net cost of fuel. In wet milling, all extraneous material is removed so only a refined starch stream is converted to ethanol. A wet miller can also divert his starch stream to make high fructose corn syrup, which is used in soft drinks, depending on market conditions. In addition, corn oil and corn gluten feed and meal are produced and sold. Larger plants also can be adjusted to burn low-cost high-sulfur coal, use cogeneration, and capture carbon dioxide for internal uses or to sell in the industrial gas market. The end result is a lower net ethanol cost.

Although wet milling is generally recognized as being more efficient than dry-mill techniques, the cost of producing ethanol varies widely from plant to plant, depending on the original decisions concerning the plant's location, process, size, energy source and whether it stands alone. With so many variables, experts that follow the ethanol business report plant product costs ranging anywhere from 94 cents to $1.73 per gallon, assuming a delivered corn cost of $2.50 per bushel.

FUTURE TRENDS IN ETHANOL PRODUCTION

Ethanol producers are optimistic they can bring ethanol production costs down to a level more competitive with gasoline within a decade. Some of the more promising research now being performed in this area is listed below.

ENZYMES

Developing highly sophisticated enzymes can accomplish several goals in reducing ethanol manufacturing expenses. Producers hope to boost current yields of 2.5 gallons per bushel to 3 gallons in five years. One kind of enzyme now being studied may break down complex sugars, which could increase ethanol yields and thus reduce net feedstock costs. Other enzymes being looked at would allow an ethanol producer to make less protein feeds and carbon dioxide so that the higher yields of ethanol will reduce net feedstock costs. Scientists also are designing enzymes capable of converting a higher portion of the corn fiber to ethanol, which will reduce the amount of lower valued byproducts such as corn gluten feed or distiller's dried grains, further decreasing production costs.

HUMAN GRADE
DISTILLERS DRIED GRAINS (DDG)

After the starch portion of the grain feedstock is used to ferment into ethanol, the remaining high-protein byproduct is then dried and used for animal feed. Revenue from feed sales offset the cost of feedstock used in ethanol production by approximately 40 percent. If this process were enhanced to produce a human grade high-protein food supplement, feedstock costs might be offset entirely. This could lower production costs by 30 cents per gallon or more.

The production of human grade DDG could increase revenue and decrease the ethanol plant's net feedstock cost. DDG prices today average $100 to $150 per ton. If human grade DDG were produced, its value could be worth several hundred dollars per ton and could supplement U.S. food exports. Today, market demand for such a product is still untested mainly because ethanol producers have typically been unable to afford the kind of risky investment needed to upgrade their feed equipment.

ENERGY CONSUMPTION REDUCTION

Ethanol producers are working to reduce energy costs in two ways: by increasing the energy efficiency of current processes through improved heat exchanger technology and by incorporating waste energy cogeneration technologies, such as the tire burning plant at ADM's Decatur, Ill., facility, which recycles steam to produce ethanol.

CELLULOSIC BIOMASS CONVERSION TO ETHANOL

Technology for the conversion of cellulosic biomass to ethanol using agriculture residues, herbaceous grasses and woody biomass, yard and garden wastes, and the biomass fraction of local landfills holds great promise.[7] Cellulosic biomass is an abundant, and in some cases free, resource in the United States. After all, it's not very likely we'll stop generating trash any time soon. And in cities where landfill space is at a premium, some companies could find it less expensive to convert waste to fuel than to haul it to a dump.

However, the technology to convert municipal waste and other cellulosic biomass to fuel is still being developed, and as a result, the cost per gallon is still quite prohibitive compared to the cost of producing gasoline at a refinery. Government and private research has reduced the conversion technology from about $3.60 per gallon in 1983 to $1.27 per gallon now. Additional technical targets have been identified that could bring the selling price down to 67 cent per gallon following aggressive research and development. Some of these technologies include using acids or enzymes to speed the breakdown of cellulose into component sugar molecules which can be fermented into ethanol. The challenge is to develop low-cost methods to convert naturally resistant celluloses, such as paper products and woody biomass, into alcohol. Government researchers are also examining the role that specially-designed bacteria can play in accelerating the process.

Policy experts say shifting ethanol feedstocks away from corn and toward cellulosic biomass is an essential component of any public policy requiring a greatly expanded ethanol supply due to environmental and economic factors. In the United States, it would be impractical and physically impossible for citizens to rely on one crop, like corn, for both food and fuel supplies. And while it is highly unlikely that ethanol could be replaced on a gallon-for-gallon basis with gasoline, the political support ethanol enjoys is expected to continue for the forseeable future, possibly in the form of mandated marketshare levels when used as a blendstock for gasoline.

DEMAND

Currently, about 10 percent of the U.S. motor fuel pool contains ethanol, with the vast majority used as E10.

Inherently, ethanol's major value should result from its high blending octane number and its oxygen content, which reduces both hydrocarbons and carbon monoxide from a car's tailpipe. However, because ethanol is confined to downstream blending with the resulting punitive pricing structures, it will probably never compete with gasoline on a gallon-for-gallon basis. Instead, its biggest potential market lies as a feedstock for petrochemicals such as ETBE. The ethanol-based ether improves octane number, adds oxygen, has low vapor pressure, and may be blended in refineries and transported by pipeline. As tighter clean fuel rules are put in place, ETBE is expected to gain further acceptance from refiners.

ENVIRONMENT

GREENHOUSE GAS

As a nonfossil transportation fuel, ethanol is often touted for its low impact on greenhouse gas emissions. Burning ethanol in a vehicle engine sets up a carbon cycle — corn plants require carbon dioxide (CO_2) for growth, so the CO_2 released in combustion is taken up again during the growth process.

There is some debate on this assessment since some argue that when considering the entire fuel cycle of ethanol on a Btu basis, one is expending more energy through natural gas or coal combustion to produce ethanol than other fuels such as petroleum. Ethanol producers disagree with this analysis, saying the energy expended to draw a barrel of crude from the ground should be considered in any assessment of greenhouse gas production. The debate is expected to rage for some time, with no clear consensus seen.

Ethanol as a pure fuel or in a blend has the ability to reduce mobile source pollution such as hydrocarbons and carbon monoxide, and air toxins such as benzene.

In the case of carbon monoxide, the EPA has stated that gasoline containing 10 percent ethanol will reduce CO exhaust emissions to a greater degree than 15 percent MTBE because ethanol has a higher oxygen content. High volume or 100 percent ethanol fuels also reduce carbon monoxide efficiently. However, ethanol fuels do appear to have some environmental disadvantages as well. Ethanol used in gasoline can exacerbate ambient levels of ground level ozone although ethanol used in concentrations of 85 percent and higher does not.

The reason for this is chemical, like anything else related to motor fuel emissions. At room temperature, low molecular weight substances like ethanol would be gases if polarity did not keep them collapsed as liquids. Polarity in a molecule is a highly cohesive molecular bonding force. When ethanol is dissolved in a nonpolar solvent like gasoline, the alcohol molecules become separated, and the molecular structure is weakened. As a result, the ethanol behaves more like a gas, resulting in an increase in vapor pressure for the gasoline mixture larger than would be expected based on the alcohol's and gasoline's pure liquid vapor pressures.[8,9]

However, this higher volatility doesn't occur in the case of high volume ethanol blends. When alcohol concentrations are 85 percent by volume or higher, the vapor pressure is lower than gasoline and therefore less likely to react to create smog.

And there is another problem. Automakers that have tested high volume ethanol fuels say aldehyde emissions reach levels that require special catalytic converters for correction. Aldehydes are not thought to contribute to smog, but they do fall under the same kind of air toxin category that formadelydes do. Future air pollution policies may tighten the emission levels of aldehydes, making an ethanol-fueled car a less attractive option from an environmental standpoint.

CONCLUSION

B ecause ethanol fuels are in liquid form like gasoline, they require the least amount of capital investment for retail gasoline marketers, but they are also usually the most expensive for motorists to purchase because of high production and distribution costs when compared to traditional fuels. For this reason, ethanol fuels are expected only to be used in niche markets, mostly supplying federal or state government fleets which are mandated to use the fuel.

References

1. University of Wisconsin. *Alternative Fuels Refueling Station Infrastructure Study*, April 1, 1994.
2. Owen, Keith and Coley, Trevor. *Automotive Fuels Handbook* (Society of Automotive Engineers, 400 Commonwealth Drive, Warrendale, Pa., 1990).
3. American Petroleum Institute. Editorial and Special Issues Department. *Federal Energy Subsidies*, August 12, 1994.
4. Lorenzetti, Maureen. "How to Get Into the Business of Supplying Alternative Fuels," *Independent Gasoline Marketing Magazine*, July/August 1994.
5. "Ethanol Excise Tax Exemption Economical, Says RFA Study," *Oxy-Fuel News*, April 17, 1995.
6. Keim, Carroll. "Why Don't Refineries Make Ethanol? *Fuel Reformulation Magazine*, March/April 1992.
7. Wyman, Dr. C.E. "Cellulose and Biomass Conversion Technology and Its Application to Ethanol Production from Corn," *Fuel Reformulation*, March/April 1993.
8. EIA. *Alternatives to Traditional Transportation Fuels*: An Overview, June 1994.
9. American Petroleum Institute. *Alcohol and Ethers*, Publication No. 4261, July 1988.

CHAPTER SEVEN

COMPRESSED AND LIQUEFIED NATURAL GAS

Fueling one's car with natural gas may seem a distant possibility to the average motorist, but more and more U.S. fleet managers are using this alternative to gasoline on a daily basis. In fact, at the present time natural gas is the fastest growing alternative fuel in the United States today, thanks to fuel price and an organized effort by several natural gas utilities to bolster a natural gas refueling network.

Today, there are close to one million in operation worldwide, with over 35,000 in the United States, primarily in fleet operations. But policymakers expect that number to increase dramatically due to several new federal regulations that have been implemented in the past year under the Clean Air Act Amendments of 1990, the National Energy Policy Act of 1992, and the Alternative Motor Fuels Act of 1988.

FUEL CHARACTERISTICS

Natural gas is composed primarily of methane (CH_4), along with other hydrocarbons, such as ethane, propane, and butane, and inert gases such as carbon dioxide, nitrogen, and helium. Its composition varies depending on location and source. Most natural gas comes from underground wells and it is almost always found in crude oil patches. Another less common but renewable source of natural gas comes from the anaerobic decomposition of animal waste and vegetable matter which produces biogas.

As a vehicular fuel, it comes in two forms: compressed or liquefied. Compressed natural gas (CNG) is a gaseous fuel that is processed for automotive use by compressing low pressure pipeline-quality natural gas to 3,000 to 5,000 psi in the storage tanks of fast-fill CNG refueling stations. It is dispensed and stored in vehicles at pressures of 2,400 to 3,600 psi. Liquefied natural gas (LNG) is gas that has been liquefied at -260 degrees Fahrenheit and stored in insulating tanks. LNG has the advantage of greater range per volume fuel, but

Photo 7-1   Optimized 5.2 Liter V-8 Engine in Dodge CNG B-Model Van (92-CTPP-1019)

is not as widely used in the United States because of its refrigeration needs. LNG is often confused with LPG, or liquefied petroleum gas, commonly called propane, which is a liquid mixture of at least 90 percent propane, 2.5 percent butane and higher hydrocarbons, and the balance ethane and propylene (see Chapter 8). It is a byproduct of natural gas processing or petroleum refining. LNG, on the other hand, is natural gas that has been cooled to liquid form.

Most gasoline or diesel-powered vehicles can be converted to operate on CNG or LNG exclusively or in combination with traditional petroleum fuels. The physical and chemical properties of natural gas provide good engine performance with no loss of power. Original manufacturer produced CNG vehicles tend to be cleaner-burning but have no significant other benefits over a converted vehicle.

Technical advantages of using natural gas include high octane rating, improved cold starting, potential reduced maintenance resulting from decreased deposits and lube oil degradation, price of fuel, and decreased automotive emissions. Some disadvantages associated with these fuels are added weight and space of fuel storage tanks; shorter driving range between fill-ups, anywhere from one-half to two-thirds the mileage of a gasoline-powered vehicle; limited availability of refueling stations; possible global warming contributions; and fuel system modifications required on retrofits.

But the impressive combination of industry-backed research, plus state and federal financial support for NGVs, may help overcome the shortcomings of today's NGVs. An industry consortium led by the DOE and Johns Hopkins University Applied Physics Laboratory (APL) has developed a prototype dedicated CNG four-door compact sedan that demonstrates a 315-mile driving range between refuelings, which is about twice the range of CNG vehicles on the road today, and a fuel economy of 32 miles per gallon. APL says 75 percent of the vehicle's original trunk space will be retained by reengineering the rear underbody for fuel storage and by compensating for the absence of a spare tire through the use of special tires which can be driven for up to 250 miles with no air in them without harming the vehicle or the wheel. For now, the cost of such a vehicle is still too expensive to be considered for the mass market, so scientists are focusing their second research phase on lowering manufacturing costs and enhancing the driving range to at least 350 miles before 1997.

Processing

Natural gas composition varies nationwide, depending on original gas composition and processing. Most of the natural gas available does not vary significantly in content, with the exception of small gas utilities where there may be large differences in hydrocarbon content.

At the crude oil/natural gas drilling site, natural gas goes through a field separator, which separates crude oil from the gas. The gas is then shipped by pipeline. Sometimes, further gas processing needs to be done because there are some heavy hydrocarbons left in the material that can condense in cold weather, disrupting shipping. Gas production companies may also have to purify their product because of gas sales contracts, which typically require the dew point to be no higher than 15 defrees Fahrenheit at a pipeline pressure of 800 psi. Natural gas destined for the LNG market is cooled to -260 degrees Fahrenheit. In these conditions, about 85 percent of the ethane and all of the heavier hydrocarbons, such as propane, drop out of the gas.

Gas compressed for CNG motor fuel comes directly from pipelines and is compressed on the spot at the refueling facility. This means the only major constraint on locating CNG refueling stations is that they be in an area with natural gas service, and today about 91 percent of the U.S. population lives in counties that are already serviced by natural gas.

Gas pressure in the pipeline depends on the pressure at which utilities store the gas in the pipelines and the distance from a pumping station. The higher the inlet pressure, the better suited the location is for a CNG refueling station. Higher pressure gas requires significantly less energy to compress it to the 3,000 psi level or higher needed for CNG vehicles. In fact, it takes almost as much energy to compress natural gas from near 0 psi to 300 psi as it does to compress it from 300 psi to its final storage pressure. Therefore, locating a CNG refueling station on a relatively high pressure natural gas line can significantly cut station operating costs.

Another important area to consider is gas composition, for two important reasons.

■ Trace amounts of sulfuric acid can form from the presence of hydrogen sulfide and water vapor in the natural gas supply. This increases the risk of CNG cylinder corrosion from the formation of sulfuric acid.

■ Large amounts of nonmethane hydrocarbons will enrich the fuel mixture, reduce octane number, and increase hydrocarbon emissions.

NGV users and suppliers should know the local composition of natural gas so that engine parameters such as air/fuel mixture and ignition timing can be adjusted accordingly. Knowing the fuel's octane and cetane ratings also helps ensure driveability. The octane rating of a natural gas fuel is just like that of a gasoline mixture in that it is a measure of its resistance to pre-ignition under compression, referred to as *knock*, or spontaneous combustion. The higher the octane number, the more resistant the fuel is to knock. Good antiknock properties allow for use of an increased compression ratio, with a resultant increase in energy efficiency. Natural gas generally has an octane number in excess of 120, compared to 87 for unleaded regular and 92 or 93 for premium gasoline. The cetane rating of a fuel is the measure of its ability to auto-ignite when compressed and heated. A high cetane number is essential for proper compression in a compression or diesel engine. Since natural gas has a very low cetane rating, it cannot be used in a diesel engine without an external ignition source such as a spark plug, a glow plug, or pilot injection of diesel fuel.

When looking at the quality of a natural gas stream destined for vehicle use, the Society of Automotive Engineers Fuels and Lubricants Technical Committee's J1616 guidelines are a good rule of thumb for proper natural gas fuel composition.

STORAGE AND DISTRIBUTION

P rimarily used for heating and cooking, natural gas is distributed through a well-established infrastructure made up of three segments. Production companies explore, drill, and extract natural gas from the ground. Transmission companies operate the pipelines linking the gas fields to major consumer areas. Distribution companies are the local utilities delivering natural gas to the customer. Natural gas is delivered to more than 174 million American consumers through a 1.2 million mile network of underground pipe, so

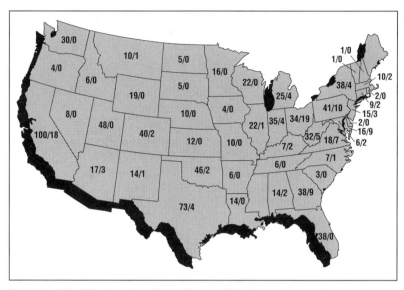

Fig. 7-1   U.S. CNG Refueling Sites.   Source: U.S. Department of Energy

unlike other alternative fuels like hydrogen, an NGV infra-structure is at least partially in place.[1] Capital investments have already been made at the wholesale level through decades-old pipelines. The next step, retail refueling stations, is growing, but is not as established as the corner service station. Refueling stations are quickly expanding, mainly though the aggressive efforts of natural gas utilities, which have been able to fund public refueling sites through ratebasing at the consumer level. Through this expansive marketing effort, more than 5 million NGVs could be on U.S. roads by the year 2010, provided there is continued government support for AFV programs. If this prediction comes true, it will mean big investments for the nation's fuel suppliers. Approximately 15,000 refueling stations and $4.3 billion will be needed to serve light-duty vehicles, and about 5,000 centrally-fueled locations will be required to meet the needs of NGV customers.

Right now, there are about 900 natural gas refueling stations in the country.[2] Most are centrally located refueling sites for fleet operators, but many refiners have opened up CNG pumps at retail service stations that also sell gasoline.

TECHNICAL CONSIDERATIONS FOR REFUELING

COMPRESSED NATURAL GAS

C NG refueling transfers natural gas under pressure and may be set up as either slow-fill or fast-fill. Slow-fill is designed for home use to be used overnight with a six- to eight-hour recharge. Fast-fill systems are practical from a motorist's standpoint because they take three to five minutes to refuel a vehicle, which is only slightly longer than gasoline refuel times.

Fast-fill refueling systems consist of a substantial amount of high pressure storage and one or more compressors capable of replenishing the stored gas supply as it is dispensed to vehicles. This type of system is capital intensive, with a commonly quoted cost for a standard 100 cubic-feet-per-minute public access station of about $250,000. Large quantities of gas must be stored since only about one-third of the storage capacity is available for refueling vehicles due to the pressure differentials between the storage cascade and the vehicle.

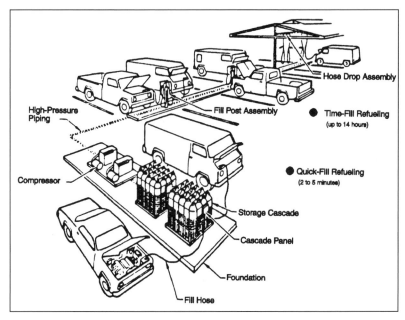

Fig. 7-2 Typical CNG Vehicle Refueling Station with Fast-Fill and Slow-Fill Systems. Source: U.S. Department of Energy

Fuel availability depends on the storage capability and the speed at which the compressors can replenish this supply as it is drained. As a result, fuel suppliers can tradeoff between the compressor size and the quantity of gas stored to provide the necessary fuel for a fleet.

Here are two typical fast-fill station configurations:[3]

1. Cascade Storage Systems — these stations are made up of small to medium (25 to 100 standard cubic feet per minute) compressors with relatively large (20,000 to 60,000 standard cubic feet) amounts of cascade gas stor-

age. Suppliers that use this system store fuel in three tanks, all at different pressures. The vehicle is fueled in a cascading sequence, in which the gas is drawn first off the lowest pressure gas storage bank, then the medium pressure, and finally from the highest pressure bank to top off the vehicle fuel tank. The compressor fills the storage banks in reverse order.

2. Direct Fill Systems — these stations generally consist of large to very large, 100 to 300 scfm, compressors with a small quantity, about 10,000 scf, of buffer gas storage. In these systems, the gas is compressed directly from the compressor into the vehicle fuel tank. The buffer gas storage serves only to augment the compressor in times of peak refueling. The cost of gas storage can be quite high, but a direct fill station is usually more efficient than a cascade system when very large quantities of CNG fuel are required.

For either system chosen, the CNG fast-fill station components generally include the following:

■ electric compressors with methanol injection or inlet/outlet gas drying equipment and blowdown tank
■ storage cascade or buffer system
■ sound-attenuating enclosure
■ gas-flow metering system
■ digital display dispensing pumps with up to two quick-disconnect hoses and one sequencer panel per dispensing pump

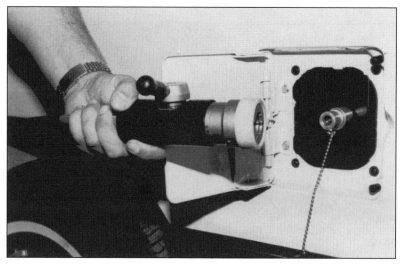

Photo 7-2   Dodge B-Model Van's Quick Fuel Filler (92-CTPP-1016)

■ explosion-proof electrical panels
■ an enclosure for the compressor and panels
■ crash posts

Slow-fill, sometimes called *time-fill* or *overnight,* stations are simpler and less expensive than fast-fill ones but require that the vehicles be stored four to 12 hours at the refueling station. Instead of compressing the gas prior to dispensing it to vehicles, a time-fill system fills the vehicles at a slow rate, using regulated street-pressure gas, compressing the gas directly into the vehicle's on-board storage. Along with being less capital intensive, slow-fill stations can be operated during night hours when electricity demand is low and the cost to operate the compressor is less expensive.

CNG slow-fill stations typically include the following:

- electric compressors with methanol injection or inlet/outlet gas drying equipment and blowdown tank
- sound-attenuating enclosure
- storage cascade (optional)
- gas-flow metering system
- simple refueling dispenser, such as a fuel post, with quick-disconnect hoses and swivels
- sequencer panel
- explosion-proof electrical panels
- time-fill hook up posts and hoses
- crash posts

In order to standardize CNG refueling sites across the country, the natural gas industry has adopted the NGV1 standard for fueling nozzles. This design was created so vehicles can be refueled nationwide without compatibility worries. In addition, the National Conference on Weights and Measures has adopted a gasoline gallon equivalent (GGE) measurement to help consumers become more familiar with the fuel. Defined as 5.66 pounds CNG, the GGE was developed to approximate the equivalent energy of a gallon of gasoline. This value was selected because vehicles will travel about the same distance on this amount of CNG as they will on a gallon of gasoline.

LIQUEFIED NATURAL GAS

LNG refueling transfers a cold liquid under pressure of around 15 psi and generally takes a few minutes longer than conventional refueling because a greater volume of liquid is transferred to compensate for the lower energy content of the fuel. LNG is dispensed like LPG with similar refueling times.

Typically a cryogenic liquid such as LNG is stored in a double-walled tank that looks like a thermos bottle. The inner tank is wrapped with a mylar insulation that can keep liquids cold for days and sometimes weeks.

LNG is usually trucked or vaporized and reinjected into the pipeline. When used as a motor fuel it can be shipped by truck. The vast majority of LNG users are heavy-duty vehicles, such as buses or locomotives that have highly specialized refueling systems. All of them tend to mimic gasoline refueling sites because of the liquid nature of the fuel.

SAFETY ISSUES

Refueling must be done in a well-ventilated area to prevent accumulation of dangerous gas levels. In addition, state and local regulations regarding NGV refueling may make it too expensive to build an indoor refueling site. The significance of this issue may diminish as local fire and building code inspectors become more knowledgeable about this fuel, but for now, there is still considerable opposition to natural gas vehicle use in tunnels or over bridges in some areas because of flammability considerations.

LNG users should be aware that severe frostbite may occur if one comes in contact with the liquefied gas or associated cold components such as storage tanks.

The industry generally has enjoyed an excellent safety record, attributed primarily to the structural integrity of the NGV fuel system and the physical qualities of natural gas as a fuel. Samples of NGV fuel cylinders have been subjected to a number of federally required severe abuse tests, such as pressure extremes, gunfire, heat extremes, collision, and fires. In all cases, the cylinders have passed with excellent marks for safety. With regard to fuel leaks, NGV fuel systems are closed, preventing any spills or evaporative losses. During those rare times that a leak does occur, the natural gas dissipates into the atmosphere because it is lighter than air. Natural gas also contains odorants to aid in detection of leaks. Natural gas is not toxic and is believed to be noncarcinogenic. Due to its gaseous form, however, it can serve as an asphyxiant, so proper ventilation is necessary in enclosed areas.

Even though the vast majority of NGVs have experienced no safety problems, recent reports of two cylinder ruptures forced automakers, the gas industry, and conversion companies to reexamine cylinder design. In both cases, the ruptures were due to corrosion of a fiber overwrap on the cylinders. Following the two incidents, General Motors bought back all 2,500 1992 and 1993 natural gas-powered pickups so it could study environmental effects on the fuel cylinders and use the materials to help advance state-of-the-art technology. Conversion companies say that the use of a stronger epoxy coating on the cylinders may alleviate corrosion generated by road salt and weather conditions.

Economics

N atural gas is a bargain at the pump compared to nearly all other motor fuels, petroleum and non-petroleum. But other parts of the equations, such as conversion costs, which can run as high as $4,000, or refueling stations, at $350,000 or higher, could dampen the enthusiasm of some fleet managers. (see Chapter 11).

SUPPLY

Natural gas is now the principal fuel for the residential, commercial, and industrial sectors, and is projected to remain a major source of energy for at least the next 25 years. With growing interest in using natural gas as a transportation fuel, concerns have been raised that its widespread use could result in fuel shortages or price increases for these traditional customers.

In general, gas supplies are abundant, and gas pipelines, for fuel transport and distribution, are extensive. Conservative estimates show that recoverable gas resources in the lower 48 states can serve current demand for another 60 to 70 years. In areas that experience extremely cold temperatures, there is a possibility that temporary gas shortages may occur. According to government and industry estimates, about 88 percent of the natural gas used in the United States and Canada is produced in the United States. The rest comes from U.S. trading partners Canada and Mexico. Natural gas interests like to point out that using North American natural gas instead of oil or other fuels imported from the Middle East and other politically sensitive areas improves U.S. energy security and the U.S. balance of trade. Industry estimates that annual supplies of available natural gas will be more than 25.2 Tcf by 2010. By comparison, natural gas consumption was 21.7 Tcf.

PRICING ISSUES

Regulatory changes in the United States have brought on a climate which promotes greater competition and lower prices. Existing and under-construction pipeline capacity, both utility and private owned, is adequate to meet any increase in demand from NGVs. Whereas the total supply and availability of gas is extensive, the sources of the gas can be expected to shift. This, however, will not have any material impact on the end user. On a gallon-for-gallon retail basis, natural gas costs about 71 cents per gallon compared to $1.07 to $1.21 for conventional gasoline. Future CNG prices are expected to remain below gasoline.

Because of price and clean-burning characteristics, many in the oil industry have begun to dabble with natural gas refueling stations. Amoco, Shell, Phillips, Texaco, Fina, Diamond Shamrock, Total, Mobil, and Unocal all offer CNG in addition to their traditional product slates. However, the cost of retailing CNG remains a major obstacle for many fuel suppliers. CNG stations that use a fast-fill technology can require a capital investment of anywhere from $350,000 to $500,000, depending on the number of refueling sites, real estate, and other factors.

DEMAND ISSUES

To understand how NGVs might impact the overall demand for natural gas, one must estimate how popular these vehicles will be in the future. Assuming that 1 million NGVs were operating in 2010, for example, they would need about 50.1 billion cubic feet of natural gas in that year. When one considers the nation's total natural gas needs, this additional demand equals about a 3 percent increase in current natural gas demand. Therefore, increased popularity of NGVs is not expected to pose a problem for other energy users.

INFRASTRUCTURE CONSIDERATIONS

Because of the huge capital cost, most natural gas refueling sites are built directly through a local natural gas utility. Often a local power company will directly target fleets with central fueling facilities and then install a compressor station at the customer's site.

Major oil companies such as Amoco are also selling their own natural gas in cities by working with a local utility. Usually the arrangement allows the utility to receive a transportation rate of about five cents per equivalent gallon delivered at the station.

Other times, utilities may prefer to install the compressor equipment at local service stations. The gas company then sells the CNG to the dealer, who profits on the fuel sale at the pump.

Still another approach for a fuel supplier is to work with a natural gas refueling company, which will lease a

piece of property from the retail station and assume the capital risk of building the refueling station.

ENVIRONMENTAL

Recent technical improvements by the auto industry,
such as fuel-injection, computerized engines, and
emission controls, have dramatically reduced emissions from gasoline-powered cars. As a result, the emissions
benefits of an NGV vehicle are not as dramatic as they once
were. Nevertheless, they still provide greater clean air benefits
than gasoline-powered vehicles. This is because natural gas is
composed chiefly of methane — a molecule made up of one
carbon atom and four hydrogen atoms. When methane is
burned, it tends to combust more completely, leaving only
carbon dioxide and water. Other fossil fuels such as gasoline
and coal have more complicated molecular structures and
this makes it more difficult to get a complete burn.

Dedicated NGVs tend to burn fuel cleaner than dual-use
or converted vehicles. But all three vehicle types are still
less polluting than gasoline engines when equipment has
been installed correctly. Dedicated NGVs have no evaporative emissions under normal operating conditions, and nonmethane hydrocarbon exhaust emissions are 40 to 60 percent less than their gasoline counterparts. Carbon monoxide levels are also dramatically cut when natural gas is used.

In the eyes of automakers and natural gas interests, natural gas vehicles can even compete with so-called zero

emission vehicles, when one considers emissions from the wellhead to the burner tip. According to research performed for the Gas Research Institute, when full-cycle emissions are considered, natural gas vehicles are as clean-burning as electric ones.

GREENHOUSE GAS EMISSIONS

While NGVs are effective in reducing carbon monoxide and ozone emissions, they may contribute to global warming because natural gas is made up of methane, a greenhouse gas. For this reason, many environmentalists consider natural gas to be a transitional fuel, to be used until a cleaner-burning technology such as fuel cells or solar energy is commercially viable.

CONCLUSION

Price, availability, and environmental considerations will bolster demand for natural gas through this decade and beyond although ultimately demand for NGVs is directly related to continued support from federal and state agencies because of the expensive investment needed in a new refueling infrastructure. From the vehicle operator's point of view, natural gas, whether compressed or liquefied, offers some big advantages. Natural gas as a vehicle fuel costs about 40 to 50 percent less than gasoline for the energy equivelent gallon. This reduces fuel cost, plus a healthy sprinkling of AFV incentives, which exist on the federal and state level in the form of tax credits for both vehicle conversions and refueling sites. Conversion costs can range between $3,000 and $4,000 per vehicle. Refueling sites can cost as much as $1 million, depending on fill technology and volume, and a typical, fast-fill station costs about $350,000.

As with nearly all nonpetroleum alternative fuels, most NGV users will be part of a centrally-fueled fleet operation which has a regular driving pattern that can take advantage of the lower fuel and maintenace costs. And as automakers expand the introduction of factory-built, original equipment manufacturer (OEM) NGVs, instead of after-market converions, the economic advantages of NGVs are improved. This may provide an expanded market opportunity for these vehicles beyond fleet applications if the cost is comparable to gasoline-powered vehicles.

References

1.   American Gas Association. *Natural Gas Vehicles, Clean Energy for the Road Ahead*, June 1995.
2.   Lorenzetti, Maureen. "Natural Gas Industry Sees Big Market For Transportation," *Natural Gas Horizons*, October 1993.
3..   University of Wisconsin. *Alternative Fuels Refueling station Infrastructure Study*, April 1, 1994.
4.   Cannon, James. "Paving the Way to Natural Gas Vehicles," *INFORM*, 1993.

CHAPTER EIGHT

LIQUEFIED
PETROLEUM GAS

Most consumers have had contact with liquefied petroleum gas (LPG) when it is used in applications for home heating, stoves, and stationary engines. The United States is one of the world's largest producers of LPG, with about 20 billion gallons produced in 1985. LPG is like natural gas in that it is a simple mixture of hydrocarbons, mainly propane/propylene (C_3) and butane/buytlene (C_4). It is supplied in four grades. In the United States, one specification, called HD-5, is the only one permitted for automotive use and contains 95 percent propane and 5 percent butane. As a result, in the United States, the terms "propane" and "liquefied petroleum gas" are used interchangeably.[1] In Europe, however, automotive grade LPG can contain more butane. The advantages of propane over gasoline and other alternative fuels include its high octane number of 104, improved cold starting, price, and potentially less mobile source pollution, including carbon monoxide and hydrocarbons. Disadvantages are limited availability of automotive-grade fuel west of the Rocky Mountains, possibility of supply shortages during periods of extremely cold weather, added weight and space of fuel storage tanks, and the cost of retrofitted engines. At the present time, the only LPG-powered OEM light-duty sedans are in Canada. U.S. fleet managers must rely on a recognized conversion system that will not void the automaker warranty (see Chapter 11).[2]

But even with its disadvantages, propane is the third most popular motor fuel in the United States, with more than 10,000 automotive grade dealers in the country. Like nearly all other alternative fuels, propane is predominately used in the fleet environment. At this writing, there are about 300,000 propane-fueled highway vehicles in the United States and more than 2.5 million worldwide.

FUEL CHARACTERISTICS

L PG is a petroleum-derived, colorless gas, typically composed of either propane, butane, or a combination of the two. About 60 percent of the LPG available in the United States is a byproduct of natural gas processing. The remainder comes from crude oil refining.

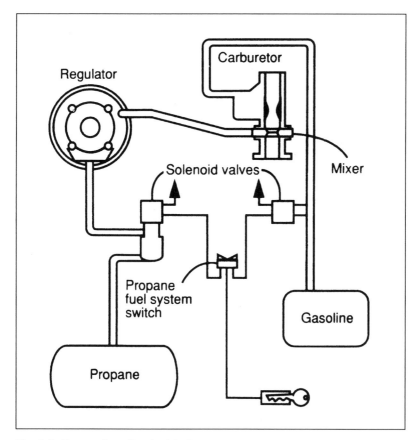

Fig. 8-1 Propane/gasoline dual-fuel system component

Refinery-produced LPG typically contains higher amounts of propylene and other low octane components than those found in natural gas-derived LPG, but the gas can be purified to meet automotive specifications.

The chemical components that make up LPG are gases at normal temperatures and pressures but will liquefy at -44 degrees Fahrenheit. Because LPG is introduced to an engine as a gas instead of a liquid that first must be atomized, it does not condense on cold manifold walls. This provides superior cold starting performance and nearly eliminates the potential for vapor lock. But at the same time, this same characteristic may limit the acceleration ability of a LPG-powered vehicle in cold conditions.[3]

The flammability range of propane, 2.4 to 9.6 percent in air, is similar to gasoline's 1 to 7.6 percent in air, while the ignition temperature of propane, 855 degrees Fahrenheit, is near the higher end of the ignition temperature for gasoline, 442 to 880 degrees Fahrenheit.

Since LPG, like CNG, is odorless, a common odorant, such as sulfur-containing mercaptans, is added to fuel grade LPG to detect leaks.

STORAGE AND DISTRIBUTION

One of the biggest advantages of propane over other alternative fuels is existing infrastructure. Propane is the most commonly sold vehicular fuel behind gasoline and diesel. Propane and other liquefied petroleum gases use a well-established pipeline and terminal storage system similar to the one in place for gasoline and natural gas although it is not as extensive as either system.

The pipeline is the main way propane is moved from production points into usage points. From the wellhead, natural gas liquids, including LPG, are extracted from natural gas in processing plants near gas producing areas. This mix of methanes, butanes, propanes, and other hydrocarbons is then moved through pipelines to fractionation facilities, which separate the different hydrocarbons into ethane, butane, propane, etc. These fractionators can be found at central pipeline hubs or large refining and chemical plant centers.[4] The fuel is then stored as a liquid on or near one of these areas and is moved by pipeline, barge, tanker, truck, or railcar to the end-user.

The propane industry counts 250 oil and gas companies producing and distributing propane, 70,000 miles of cross-country pipeline, 26,000 transport delivery vehicles, 22,000 rail cars, 60 barges and tankers, and 9,000 bulk storage distribution points. From a fuel user standpoint, the most important part of this infrastructure is the 25,000 retail outlets, with about two-thirds located east of the Rockies. However, only about half of these retail suppliers are equipped to handle transportation refueling.[5]

Still, the ability of a retail propane supplier to upgrade to an automotive refueling station is not as large an investment as other fuels. Given the existing infrastructure and the ability of the fuel to be stored before use, it costs significantly less to build an LPG refueling station compared to a CNG one.

LPG vehicle fueling stations can be operated directly by LPG supply companies, while others are run by traditional gasoline station owners. Most propane users have received training from a propane supplier to self-fuel their vehicles, with the procedure generally controlled via a cardlock system. Propane dispensing is as fast as gasoline refueling because the fuel is handled in a liquid state. Typical pumping time for a vehicle with a 60-gallon tank is three to five minutes. Propane refueling equipment looks similar to other liquid fuel systems and can be used in a cardlock fueling system to prevent misfueling. Since LPG is stored under pressure both inside the vehicle and in the refueling tank, a special pressure-resistant hose filling connection is needed to transfer the pressurized liquid from the pump to the car tank.[6]

Standards for LPG installations were first introduced in the 1930s. Since that time, standards and codes covering such facilities have been refined to increase safety and to reflect advances in the technology. The American Society for Testing and Materials, the National Fire Protection Association, and several states require tanks designed for transportation use be equipped with stop-fill devices.[7]

At the wholesale and retail level, propane is an attractive alternative fuel to supply, even with the special hose equipment, because the investment cost can be $20,000 less than a typical gasoline station and the fuel is easy to obtain. Because the fuel has a lower energy content than gasoline, propane marketers need to store larger quantities of fuel than do gasoline or diesel stations servicing a similar fleet. Most refueling systems employ 500 to 1,000 gallon storage tanks, but storage of up to 30,000 gallons is not uncommon.

In many states, refueling installations must go through a pre-installation plan review and approval process. Wellhead separator permits may be required, and nearly all local areas require building permits. It's also a good idea to check with the local fire inspector.

Propane refueling stations are completely pressurized systems. Pressurized components include the bulk fuel storage tank, dispenser, nozzle, and vehicle fuel tank. Because LPG has a relatively mature refueling infrastructure, nozzles and other refueling station components are fairly standardized. Each component must meet Occupational Safety and Health Administration specifications for pressurized vehicles. Station components include a pressurized above-ground fuel storage tank, with a single-walled underground tank optional in many areas; concrete pad; support piers for above-ground tanks above 1,000 gallons; digital display pressurized dispensing pumps with a vapor recovery system and break-away couplings; dispenser islands and safety crash posts; installation equipment such as electrical wiring; optional auto keycard access system; fuel tracking console and printer; optional canopy and lights. Typically, propane is stored above ground.

SAFETY AND ENVIRONMENTAL ISSUES

L PG is a nontoxic gas. High LPG concentrations reduce oxygen levels and may cause asphyxiation, with early symptoms of dizziness. No harmful longterm effects have been reported from exposure to propane vapors. An odorant added to LPG generally enables its detection at concentrations below the lower flammability limit and much below the concentrations needed for asphyxiation.

Because propane exists in both a liquid and gaseous phase during storage, care should be taken to allow enough room for vapor to accumulate in the storage tank. Most propane tanks, whether bulk storage, vehicle, or tractor-trailer distribution tanks, should not be filled with liquid propane in excess of 80 percent of the rated liquid capacity of the tank. Recommended fire and safety precautions include prohibition of smoking and naked lights, and that electrical equipment and motors be explosion proof. Grounding is required for all equipment, including tanks, pipe racks, pumps, vessels, and filters. In most areas, fire codes prevent self-service operation. However, cardlock systems may allow unattended fueling, depending on local jurisdiction.

LPG is not a cryogen, and liquid temperatures of the fuel at tank pressure remain at ambient levels. However, the rapid evaporation of the fuel at atmospheric pressures can, if spilled, cause damage to skin. To avoid direct propane contact to the skin, it is recommended that gloves be used during refueling.

Propane has a narrow range of flammability compared to the other transportation fuels. The fuel will only burn within a fuel-to-air ratio of 2.2 percent to 9.6 percent. Propane will rapidly dissipate beyond its flammability range in the open atmosphere. It is important that garages housing gaseous-fueled vehicles be properly ventilated. LPG fuel leaks can pose a significant explosion hazard, relative to gasoline, in enclosed garages. All forms of combustion within these enclosed spaces should be eliminated.

Because propane spills dissipate into the air, the fuel does not cause environmental contamination and does not require environmental remediation. As a result, there is no need for leak detection equipment or spill dikes, although fire protection is required due to the high vapor flammability.

Fueling systems are not required to have a fire protection system, but a fire extinguisher must be located within 10 feet, and an emergency shut-off switch within a zone of 25 to 75 feet from the dispenser.

The NFPA publishes a model code known as Pamphlet No. 58, which has been adopted by all but three states in the United States. It is the basis of standards for the Uniform Fire Code and is updated on a three-year cycle. NFPA 58 is the most complete code of its kind, detailing all facets and safety requirements for the installation of propane systems for refueling and installation of equipment on vehicles.

ECONOMICS

SUPPLY AND DEMAND

LPG is a fossil fuel byproduct of either natural gas processing or crude oil refining. When natural gas is produced, it contains methane and other light hydrocarbons which are separated at a gas processing plant through a combination of increasing pressure and decreasing temperature. The natural gas components recovered during this process include ethane, propane, and butane and heavier hydrocarbons. The amount of propane varies widely among different natural gas streams.

Propane is also produced during crude refining as a byproduct of refinery techniques used to rearrange or break down the molecular structure of crude into more desirable compounds such as gasoline or petrochemicals. Reforming and catalytic cracking, for example, are two refinery processes that are the primary sources of propane, butane, and ethane. Ethane is typically recycled within the refinery as a plant fuel, but propane and butanes are a mixture of paraffins and olefins, which are building blocks for a number of refinery products, including gasoline and diesel fuel. Butane is not used directly as a motor fuel in the United States, but 20 billion gallons per year of the hydrocarbon and related refinery-produced hydrocarbons, including propylene and butylenes, are used in the production of gasoline. N-butane is blended into gasoline to provide volatility in cold weather conditions, while isobutane is used with other refinery streams to produce gasoline components such as alkylate and ethers.[8]

In the United States, gas plants currently supply about 13 billion gallons of LPG, about 60 percent of the LPG marketed in the United States, excluding LPG produced and used captively in refineries. According to industry and government estimates, gas plant production will not change significantly in the next decade since increases in gas volumes processed will be offset by reduced gas liquid content. By 2004, the availability of gas plant propane is estimated to be 7.7 billion gallons.

Refinery-generated LPG, meanwhile, is steadily increasing due to higher conversions to gasoline, increased severity of refining processes required to meet the demand for reformulated gasoline, and the processing of heavier crudes. Refiners market propane output not used in other refinery streams. By 2004, refinery-based propane output is expected to reach 8 billion gallons.

Imports, mainly from Canada via pipeline and rail, also make up the supply picture. Some propane is also supplied through overseas markets, including the Middle East. About 30 percent of U.S. LPG sales are from imports. Propane suppliers say that U.S. and Canadian sources of LPG are sufficient to support an expanded propane vehicle market of 3.5 million vehicles within 10 years.

The largest single demand for propane, as with natural gas, is for residential and commercial uses. In 1994, 37 percent of U.S. propane consumption was for residential and commercial heating. Propane is often substituted or supplemented with natural gas for peak shaving, space heating, cooking, and clothes drying. It can also be used as a petrochemical feedstock, depending on market conditions.

Motor fuel use accounts for about 5 percent of demand now. But the DOE estimates that if the government's goal of replacing 30 percent of U.S. motor fuels with alternative fuels is achieved by the year 2010, propane demand for vehicle use will rise from today's 35,000 b/d to about 1.7 million bld, a fifty-fold increase.

Since most residential propane users do not have access to less expensive natural gas, they have limited choice in fuels. As a result, there is some concern increased demand for nontransportation propane will mean higher prices for residential, agricultural, and industrial customers.

There also is the issue of supply source. Government estimates say any increase in U.S. demand for propane will have to be met through imports. According to U.S. Department of Commerce data, between 1989 and 1994 propane imports rose by 159 percent. Right now, the majority of those imports are from Canada, but if propane vehicles become popular, some are worried that an increasing amount of fuel will have to be imported from the Middle East. But propane suppliers stress that non-Middle East sources of propane are readily available to meet future transportation needs.

PRICING CONSIDERATIONS

Since propane and butane are oil and gas processing byproducts, LPG transportation fuel is generally competitively priced relative to gasoline, but a wider variation in price can be experienced, depending upon the pricing practices of the local station operator.

Distributors in a given market may be competing with natural gas, fuel oil, and electricity in heating markets; naphtha, heavy oil, and ethane in the petrochemical market; and with CNG, diesel, and gasoline in the transportation fuel market — depending on state, federal, and local taxes and supply considerations. When adjusted on a gasoline equivalent basis, the federal highway motor fuel tax for propane is 24.9 cents per gallon compared to 18.4 cents for gasoline and

21 cents for diesel. State tax incentives and/or exceptions, however, may make a propane fuel competitive or less expensive than gasoline or diesel. Additionally, fleet customers are likely to negotiate better pricing from a fuel supplier for LPG, gasoline, ethanol, or diesel because of volume considerations.

Another important factor in considering propane is the cost of a converted vehicle, since at this writing automakers do not offer light-duty propane-powered vehicles right out of the factory, although automakers provide information on after-market conversions for specific engine types. In the case of heavy-duty vehicles, such as transit buses and trucks, some factory-made, propane-powered vehicles are expected to appear shortly.

Converting a light-duty vehicle to run on propane will cost anywhere from $1,700 to $2,500, depending on the conversion system and engine type. While that may seem expensive, it's still about half the price of converting an engine to CNG. That's because propane, unlike CNG, doesn't require the special tanks required for a high-pressure fuel. Propane's energy density is also much closer to gasoline, so additional tanks don't need to be installed, which can increase the weight of a vehicle and reduce fuel economy.

EMISSIONS

L PG was first touted as an environmental fuel in the 1950s, when it was discovered that its use in indoor vehicular equipment, such as fork lifts, reduced the risk of carbon monoxide poisoning in poorly ventilated areas. Since then, its use in the transportation sector has risen because a properly-converted system minimizes evaporative emissions, which make up 35 percent of the total vehicle emissions in the United States. Propane tailpipe emissions, which are less reactive in sunlight, have up to a 45 percent lower smog forming potential than exhaust emissions from gasoline.

A significant advantage of LPG over other alternative fuels, such as methanol, is that propane is stored and transferred under pressure in a sealed system. Therefore, losses at refueling are significantly lower, and stationary and evaporative running losses are eliminated. Compared to a low vapor pressure fuel, such as gasoline in average summer conditions, propane provides a 50 percent reduction in hydrocarbon emissions. When used in heavy-duty vehicles as a replacement for diesel, particulate emissions are eliminated. With regard to ground-level ozone, propane is about half as reactive as gasoline and is similar to methanol, except that LPG does not emit the higher formaldehyde levels alcohol fuels do.

The fact that even a properly-designed propane system emits the same levels of NO_x as gasoline is a sore spot among some environmental groups and policymakers, who say propane does not offer significant ozone-reducing benefits. Since two-thirds of the LPG consumed in the United States comes from refineries, imports, or natural gas associated with crude oil production, propane should be considered in the same class as gasoline and diesel and not receive any special vehicle tax breaks that other alternative fuels enjoy under federal law. Despite these criticisms, propane is considered an alternative fuel under both the Clean Air Act of 1990 and the Energy Policy Act of 1992.

GREENHOUSE GAS

Ironically, even though LPG is technically a fossil fuel, government reports have found that carbon dioxide emissions are lowest for CNG and LPG. Researchers said neither fuel requires much conversion or processing before it is used as a motor fuel.[9] Other fuels, particularly ethanol, contain feedstocks that require high levels of energy, and hence potential emissions, before they can be used. Also, both CNG and LPG have a very low carbon-to-hydrogen ratio to other fuels. As Dr. Mark DeLuchi of Argonne National Laboratories notes, both these fuels contain more hydrogen atoms per carbon atom than do other fuels, and a unit weight of hydrogen produces more energy than a unit weight of carbon. Both CNG and LPG produce less of the carbon dioxide greenhouse gas than other fuels as a percentage of energy consumed.

However, as fossil fuels, once they are burned, the carbon dioxide cannot be recaptured. In the case of a renewable fuels such as corn-based ethanol or biomass-derived methanol, new crops can be grown that will absorb greenhouse gases.

CONCLUSION

Propane is a readily available, nontoxic fuel that is price competitive with gasoline and enjoys a mature refueling infrastructure. It is not as environmentally friendly as electric-powered vehicles but will provide ample air pollution reduction benefits to qualify for clean fuel programs. One possible future constraint on propane vehicles is supply. There is not enough in the United States to meet an expanded market, and using imports to replace other imported fuels, like gasoline and diesel, may eventually strip propane of its government support. Another barrier to expanded propane use is the lack of OEM vehicles. Without the active participation of the auto industry in the use of propane as a transportation fuel, consumer acceptance of the vehicle may be limited to fleet use only.

References

1. Lorenzetti, Maureen. "How to Get Into the Business of Supplying Alternative Fuels," *Independent Gasoline Marketing Magazine*, August/September 1994.
2. National Propane Gas Association. *An Assessment of Propane as an Alternative Transportation Fuel in the United States.*
3. U.S. Department of Energy. *Facts about CNG and LPG Conversion*, Document # DOE/CH100093-315, DE94006888.
4. Petroleum Equipment Institute. *Alternative Fuels Refueling Equipment Technical Conference and Trade Show*, Orlando, Fla., March 16-17 1994.
5. EIA. *Alternatives to Traditional Transportation Fuels: An Overview*, June 1994.
6. *University of Wisconsin Alternative Fuels Refueling Station Infrastructure Study*, April 1, 1994.
7. Gushee, David. *Commercial Fleet Manager's Guide to Alternative Fuels*, Texas Edition, R.W. Webb Corporation, Washington, D.C., 1994.
8. Childress, Jim. *Propane Consumers Coalition Congressional Testimony to U.S. House of Representatives on Alternative Fuels Mandates and the Energy Policy Act of 1992*, July 1995.
9. DeLuchi, Mark. *Emissions of Greenhouse Gases from the Use of Transportation Fuels and Electricity*, ANL/ESD/TM-22, Argonne National Laboratory.

CHAPTER NINE

ELECTRICITY

Today's EVs are becoming a lot more sophisticated than the glorified golf carts they once were, but there are three primary barriers that must be overcome before they can compete with gasoline-powered vehicles. These barriers are range, performance, and cost. Ongoing battery research between the federal government and automakers is helping EVs become more commercially attractive, but automotive experts are uncertain the technology will be ready to meet mandates scheduled for California and possibly the Northeast in the late 1990s. Despite the considerable technical challenges associated with EVs and the need to fully develop a recharging infrastructure, the

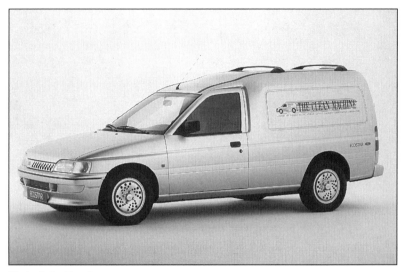

Photo 9-1   Ecostar

substantial environmental benefits outweigh the inconveniences in the minds of some policymakers. It remains to be seen whether those charged with building and driving these vehicles can be convinced that the economic tradeoffs are worth the environmental payoff.[1]

FUEL CHARACTERISTICS

Currently, the power source for an electric-powered vehicle comes from a battery, either a variation of the traditional lead-acid variety, or through new chemical compositions now being considered by automakers, including nickel cadmium, nickel iron, nickel zinc, zinc bromide, sodium sulfur, sodium nickel chloride, lithium, nickel metal hydride, zinc air, and aluminum air.

Photo 9-2   Nissan's new quick-charge battery

When considering the various available alternative fuels, battery-generated electricity is unique because mechanical power needed to power the vehicle is immediately supplied to the motor, where in the case of other transportation fuels, chemical energy through the combustion process must first be converted to supply power. Simply put, a battery is an electrochemical system just as an engine is a mechanical system. In the case of an engine, fuel is typically supplied by energy-rich compound hydrocarbons, which could be gasoline, natural gas, alcohols, or propane. In the case of a battery, the fuel is supplied by a metal located at an electrode, and the oxygen used to combust the fuel is in the form of an oxide in the opposite electrode. When the battery is discharged, its fuel is depleted, with the metal of one electrode oxidized and the oxide of the other electrode reduced.

Another way to generate electricity to power the drive train is through electrical current supplied from a fuel cell.[2] This technology, which does not require recharging, is being examined for the transportation sector although today it is commercially immature. Fuel cells are unlike internal combustion engines (ICE), turbines, and other heat engines in three important ways: Fuel cells produce power without chemical combustion, so they are inherently cleaner than ICEs could ever be; fuel cells are not subject to the same fundamental laws of thermodynamics that limit the maximum efficiency of turbines and ICEs; and fuel cells have no moving parts so there is less maintenance and noise compared to a traditional gasoline-powered vehicle.

Also under development are hybrid electric vehicles that attempt to achieve the maximum level of air quality benefits while still providing the range of a gasoline-powered engine. Most prototypes being considered today run on electricity supplied by a battery and include a small gasoline engine to recharge the battery, thereby increasing the vehicle range.[3] Emission levels from this type of vehicle still include some tailpipe emissions, but they are still significantly lower than other alternative fuels or ultra-clean gasoline. The use of an electric motor in conjunction with a small internal combustion engine reduces tailpipe emissions 70 percent compared to a gasoline engine. But an all-electric car is 95 percent cleaner than a gasoline-powered car. Hybrid vehicles powered by either a fuel cell or a battery represent a promising option for future transportation needs since they conceivably offer both environmental acceptability and the possibility of rapid refueling for extended range. The technology gap, however, is still wide compared to the internal combustion engine in terms of weight, driveability, and, most important, cost.

INFRASTRUCTURE ISSUES

With various state laws expected to mandate or strongly encourage the expanded use of EVs, both supporters and opponents of zero-emission-vehicle programs point to refueling as one of the key issues that will influence the success or failure of these vehicles.[4]

The infrastructure requirements of EVs are similar to those for any fuel — namely, convenient and affordable refueling or recharging, standardization of hardware throughout the distribution system, availability of parts, and user safety projections. U.S. automakers, government regulators, and utilities are working together to address these issues, but a standardized, nationwide recharging infrastructure does not yet exist, although California and Massachusetts have developed a rudimentary network of recharging sites.

DISTRIBUTION

Technically, 98 percent of the electric vehicle charging infrastructure already exists via the nation's power grid. The existing electric generation, transmission, and distribution system in the United States should be able to meet the needs of hundreds of thousands of electric vehicles over the next decade. This is assuming that most vehicles will be recharged at night, during off-peak hours, so electric utilities can avoid building new facilities and make the most efficient use of existing capacity. According to a 1990 DOE study, if 43 million electric vehicles were on U.S. roads by the year 2000, the national electrical supply could absorb the new market, provided that the majority of recharging occurred overnight so load management could optimize recharging. The study found that 43 million vehicles, using less efficient, 1990 EV technology, would require 797 gigawatt-hours (GWh) per day for recharging, according to research by the Edison Electric Institute. If one assumes that national demand is about 900 GWh/day, a total recharging load for that demand could fit into the late-night valley typical of daily electrical demand.

Eventually, as market penetration of EVs accelerates, upgrades and additional capacity will be needed at the local distribution level since there will be more recharging during peak periods. There are also many unanswered questions about how utilities will decide to manage this new load on their systems. What is the potential impact on a utility if all of the EVs in a given area turned on at once? Will EV owners want to charge their vehicles during the day to extend the range of travel over the incentive of lower rates? Can an EV customer get reimbursed if his car is capable of feeding its stored energy back into the utility grid? Would the utility be able to physically control the load or will this be left up to the customer? Will a customer be motivated enough by incentives to shift energy use to off-peak times?

California utilities plan to encourage off-peak charging through the deployment of time-of-use (TOU) rates that make it less expensive to charge EVs during off-peak periods, typically overnight, and more expensive to charge during peak hours. Using TOU rates will require installation of a separate meter for billing purposes. In the summer time, the on-peak rates are seven to eight times more expensive than the off-peak rates. One idea being considered by some utilities is to offer TOU rate schedules for customers who elect to forego their access to charging during peak hours. The hours that define on- and off-peak time differ by utility, season, and sector. On average, off-peak hours comprise about 33 percent of a day.

In California, which has the most established EV infrastructure to date, electricity rates being proposed range from 4 to 6 cents per kilowatt/hour (kWh) for off-peak charging to more than 35 cents per kWh in peak periods. Electricity is expected to be the least cost transportation fuel on a cost-per-mile basis — as long as charging occurs predominantly during off-peak hours.

EVs are expected to be recharged primarily at private locations, such as residential or company garages. Even though EVs have limited driving ranges, California officials say the availability of public charging facilities for full or partial recharges away from the home base — referred to as opportunity charging — will help build consumer confidence and increase the use of EVs. Likely locations for opportunity charging include parking facilities at shopping centers, the workplace, park-and-ride lots, and airports. Fleet or commercial users may also need access to public charging facilities.

POWER QUALITY

With the introduction of energy efficient devices and other nonlinear loads, a growing concern for utilities is maintaining the quality of power on the grid. As John Kennedy of Georgia Power Company notes, EVs are considered a nonlinear load which can create power quality problems into the electric grid they are connected with. Voltage flickers and electromagnetic interference (EMI) may mean TV and radio signals are disrupted and possible damage to the car's electrical system. Car companies are trying to anticipate the quality of the power the motorist will receive, which may impact the design of the vehicle. Along with design considerations, car makers are working with utilities to anticipate and correct any power quality problems before EVs are sold on a mass-production level. However, given the expected limited introduction of these vehicles over the next few years, it may take several years before load management and power quality concerns will be fully addressed.

CHARGING

Since EVs are not yet available on a widespread basis, the technology for EVs is evolving. The technology for overnight, or slow, EV recharging has already been identified and can be implemented for commercial facilities or at-home recharging. Battery packs may be recharged by plugging into either a 120- or 240-volt outlet. Both of these voltages usually are available from the local utility providing residential service.

An alternative to plugging into an electric outlet to recharge batteries is inductive charging, through which the batteries are recharged without a direct electric connection between the vehicle and the recharging system, using a paddle-like magnetic coupling device. The equipment may be permanently attached to the vehicle, permanently attached to the area where the vehicle is typically parked, or completely removable. Charging stations are being developed to be configured for both slow, 120 to 240 volt, recharges and fast, more than 240 volts. Emerging battery technology has now focused on the need for fast charging stations since some batteries can receive a 40 percent recharge, equivalent to 50 to 90 miles of range extension, in six to 12 minutes. To standardize recharging sites, the National Electric Vehicle Infrastructure Working Council (IWC) has established three levels of charging:[5]

■ Level 1: Charging that can be done from a standard, grounded, 120-volt, 3-prong outlet available at all home

■ Level 2: Charging at home or public stations functioning at 240-volt/40-amp service with special consumer features to make it easy and convenient to plug in and charge EVs at home or at an EV charging station on a daily basis

■ Level 3: A high-powered charging technology currently under development that will provide a charge in five to 10 minutes, making it analogous to filling the tank of an internal combustion engine at a local gasoline station

Of the three charging levels established, Level 2, a 240-volt/40-amp circuit, is expected to be the consumers' preference at both private and public facilities. Operating at a rate up to five times faster than Level 1, Level 2 will meet the typical driver's daily needs in three to five hours of charging — at home, work, or public charging facilities. Level 3 is not expected to become the preferred recharging system due to concerns that it may occur during the peak hours for electricity use. This could cause costly distribution upgrades and possible electricity capacity problems. Another concern with high voltage recharging stations is the overall effect on the life of the battery. It is not known how different battery types will react to high-energy recharging. In some cases, rapid recharging may cause overheating or shortened battery life.

HEALTH AND SAFETY ISSUES

BUILDING, FIRE AND ELECTRICAL CODES

EV charging facilities must meet existing electrical, fire, and building codes. The IWC is working to revise national codes to adequately address the safety needs of EV charging sites without creating overly burdensome restrictions that would unduly increase installation costs.[6] At present, the IWC continues to provide technical support to the National Fire Protection Association (NFPA) for upcoming editions of NFPA 70: National Electrical Code (NEC). Suggested revisions include the following:

■ **New Article 625 — Electric vehicle charging system equipment** • The intent of this proposed article is to provide necessary safety requirements for EV charging in one convenient location within the 1996 NEC. In addition to covering the requirements for nonhazardous, or nonclassified, locations, the proposed new article cross-references articles 500 through 516 for their requirements when EV charging occurs in a hazardous, or classified, location.

■ **Revision of Section 511-9 — Commercial garages — EV charging** • These revisions delete general requirements that apply equally to nonhazardous and hazardous locations and cross-reference the general requirements of the proposed new article 625.

The IWC is also working with Underwriters Laboratories to assess methods for personnel protection against electric shock at higher than 150 volts-to-ground, at frequencies other than 60 Hz, and for all DC systems. The project is the first step toward affordable, reliable, ground fault personnel protection for Level 3 and for some Level 2 charging rates. Key issues include elimination of hazards or hazardous situations during charging, necessary ventilation requirements, and the definition of EV fueling.[7]

The Uniform Code serves as the model building code for California, but revising the model code is a lengthy process which occurs over a four-year cycle. In an effort to have EV-ready codes in place prior to 1998, California has initiated its own building code amendment project consistent with IWC efforts. This effort is moving forward in two phases. Initial amendments to California Title 24 of the Health and Safety Building Codes are expected to be adopted in 1995 through an emergency code change process. The second phase is expected to be complete in 1997 and will use the regular rule-making process to establish more comprehensive and thorough building codes, again developed in coordination with IWC.

Local jurisdictions have authority over the enforcement of codes and can apply and amend the state codes to fit local conditions. For information on codes that apply to a specific EV charging station installation, local building and electrical code officials or the Office of the State Fire Marshal should be contacted.[8]

CONSUMER SAFETY

Given that EVs are a technology with obvious differences from conventional gasoline-powered engines, automakers and electric suppliers have been forced to address a number of safety issues. As with conventional vehicles, EVs will be expected to have full Federal Motor Vehicle Safety Standards certification or to meet all of the safety standards of conventional vehicles. Batteries will usually be enclosed and away from the passenger compartment of the vehicle to address concerns about the possible presence of flammable, toxic, or corrosive materials. Still, there is also a chance of acid leakage with flooded lead-acid batteries. Acid damage can be avoided by periodically checking batteries for leakage. Original equipment manufactured EVs are expected to use advanced lead-acid batteries or newer batteries such as nickel metal hydride. Advanced lead-acid batteries use a paste or gel rather than a liquid acid and are sealed, further making them less likely to spill.[9]

Education about electrical dangers during charging and training of personnel in commercial repair garages and service centers is also being addressed. Hydrogen, a nontoxic but explosive gas, is emitted from some types of batteries during charging. Since hydrogen is lighter than air, it will dissipate rapidly if charging takes place outside or in well-ventilated garages. EV building codes will ensure adequate ventilation. No lit cigarettes, open flames, or sparks should be present in areas of charging batteries that emit gas.

EMERGENCY RESPONSE TRAINING FOR LOCAL OFFICIALS

Ford, General Motors, and Chrysler have jointly developed a video to inform fire rescue personnel of the safety precautions to be aware of when dealing with an EV. To the extent that small manufacturers develop vehicles, they will also need to keep fire officials informed regarding the attributes of their vehicles.

In California, the state energy commission is working with the California State Fire Marshal, the utility companies, and other state agencies to develop a training program for emergency response personnel. The program will institutionalize training for firefighters and other emergency personnel on procedures for safely handling an EV in an emergency situation. Initiating training early is necessary to prepare for the widespread introduction of EVs, but further vehicle standardization is needed to enhance safety and standardize response procedures. Examples of vehicle standardization that still need to be developed include establishing a uniform power shut-off system and location and creating a mechanism to clearly identify the battery type. As the market develops and EV safety features become standardized, continued efforts will be needed to train emergency personnel.[10]

AVAILABILITY OF TECHNICIAN TRAINING

In California, policymakers have instructed that any vehicle to be sold by a major original equipment manufacturer will be fully backed by the manufacturer. The auto companies will train technicians as part of their normal operations and maintain adequate parts inventory and service locations to fully support the vehicles in the marketplace. This support includes warranties comparable to those for conventional vehicles. Many junior colleges and private training organizations are expanding their curriculum to include training for alternative fuel vehicle technologies.

MARKETABILITY

COST

The cost of buying an electric vehicle today varies widely, with price ranges from $10,000 to $100,000, just like gasoline powered vehicles. However, an EV version of a typically gasoline-powered car is still expected to cost an additional $5,000 or more because of limited production numbers and the expense of new technology. Utilities claim that if EVs were mass produced in volumes of 20,000 per year — the minimum number California is seeking to introduce later in the decade — the EV body and chassis will cost no more than their conventional vehicle counterparts. Thus, any cost difference between the EV and a conventional vehicle would be determined by the propulsion system, including drive trains, and the power source. These systems in the electric vehicle at mass production volumes will be inherently less expensive because they are simpler machines.

Automakers, however, say that even if the purchase price of the vehicles were comparable with gasoline-powered engines, which Detroit has deemed impossible because of battery considerations, consumers will not want to buy a vehicle that has limitations in range, power, and recharging, along with the expense of replacing battery packs. Federal tax credits, equal to 10 percent of the cost of the vehicle up to $4,000, are available for motorists but are not enough to generate enough consumer demand for an EV industry this decade, automakers say.

Still, the case for EV economics is strengthened some-what when life-cycle costs are considered. Fuel costs per mile for electric vehicles are estimated at 1.25 cents per mile, compared to five cents per mile for conventional vehi-cles. Additionally, maintenance costs are expected to be about one-half of a conventional vehicle because of fewer moving parts in an EV engine: EVs do not require oil changes or filters, emission control system, muffler, tailpipe, air filter, fuel filter, ignition system, or transmission. Since an EV does not need any of the routine maintenance associ-ated with an internal combustion engine, the life of the vehicle is expected to be three times that of most conven-tional vehicles, according to the DOE.

RANGE

Current battery technology creates serious limitations in range and performance that automakers are seeking to over-come in the 1990s. Today's EVs have a purported driving range of up to 100 miles of typical urban driving. However, the range can fluctuate substantially from variations in dri-ving speed and the use of heating and air conditioning, according to the Electric Transportation Coalition. From now until the latter part of the 1990s, lead-acid batteries will likely be the battery of choice for most electric vehicles. Advantages of this type of battery are many. It is based on established technology and is the lowest price alternative available today. It has an adequate cycle life, is maintenance free, and there is a recycling system in place. Still, there are significant disadvantages which have helped spur billions of private and public research dollars to new battery technology. Lead-acid batteries have a low specific energy and energy density, which increase weight and volume and decrease range. Power in a lead-acid battery also decreases as the battery discharges, and a decreased capacity occurs in colder temperatures. Finally, lead-acid batteries do not appear to be conducive to fast charging, according to the results of the Massachusetts EV Demonstration Program.

There are at least half a dozen other battery technologies under development. All are much more expensive than lead-acid, and some are more flammable or toxic. The nickel cadmium battery has a high power ratio for acceleration and a long cycle life that would reduce the high cost of repeatedly changing worn batteries. But nickel and cadmium are very expensive materials, which drives up the cost. In addition, cadmium is very toxic, and no recycling program has been established for the compound. One battery that is gaining interest with automakers is nickel metal hydride, which has high power and very high specific energy and energy density, allowing greater range. A big problem, though, is again the cost of the materials, and sensitivity to high temperatures. Still, this battery type is growing in popularity, and there is a possibility that the economies of scale may lower the price of materials in the near future.

Aside from batteries and fuel cells, some companies are considering mechanical batteries, or flywheels, based on the same principle as a potter's wheel. Electricity is used to spin a flywheel at high speeds. To operate the car, the flywheel slows as power is drawn. Using this technique, a flywheel could power a car for 300 miles and be fully recharged in 15 minutes. But as with other battery technology, automakers have not found a commercially-acceptable application for an EV, although research is continuing.

When considering EV technology, a motorist should consider the following battery criteria:

■ energy density (Wh/kg), determines the driving range
■ power density (W/kg), measures acceleration
■ cycle life, total mileage of battery
■ calendar life, how long the battery will last whether used or not
■ recharge time, as fast as possible
■ replacement cost
■ recyclability and environmental toxicity

The following is a brief description of near-term battery technology, which is limited to lead-acid, nickel iron, nickel cadmium, and sodium sulfur. A brief desciption of emerging technologies such as metal air and lithium-aluminum is also included. Other compounds being considered farther down the road by automakers include zinc bromide, zinc nickel oxide, lithium polymer electrolyte, and sodium metal chloride. Each of these designs has its own performance characteristics, maintenance, and safety issues.

LEAD-ACID

Lead-acid batteries are a proven, mature technology since they are already in place in conventional light- and heavy-duty vehicles. Lead dioxide is on the positive electrode, with metallic lead on the negative electrode. The charge-discharge process is stable. When a lead-acid battery reaches its full charge, it produces hydrogen and oxygen, causing water loss. As a result, open lead-acid systems require periodic watering, about once a month. Automakers, however, are moving to maintenance-free, sealed-cell, lead-acid batteries that use a gel electrolyte or absorptive glass matte. Lead-acid batteries are the least expensive to buy and operate since they have a relatively long life span, up to three years in current EVs. They can operate in a wide range of temperatures, generally from 0 to 40 degrees Celsius, and provide moderately high power density. At room temperature, lead-acid batteries designed specifically for an EV will self-discharge between 4 and 6 percent per month. But there are significant disadvantages that could limit their long-term use in EVs. These include excessively deep discharge, weight, and cold weather operation degradation. Also, in the case of an EV, excessively deep discharge can permanently damage the battery, as will storage when the system is uncharged or partially-charged.

NICKEL CADMIUM

Nickel-cadmium batteries (Ni-Cd) use cadmium on the negative electrode and nickel oxide on the positive electrode, coupled with a potassium hydroxide solution electrolyte to spur a reaction. This kind of battery is considered mature technology and is commonly used for a variety of consumer products such as cellular phones, notebook computers, and power tools. Ni-Cds can operate in a wider temperature range than lead-acid, -40 to 50 degrees Celsius, with high power density. At room temperature they self-discharge between 10 and 15 percent per month. Sealed Ni-Cds batteries are maintenance-free and offer good, high-rate/low-temperature performance. Disadvantages of this technology for the EV market, however, remain significant. Ni-Cd batteries cost four times as much as lead-acid. Also, from an environmental standpoint, a new recycling/disposal system would have to be developed because of the highly toxic nature of cadmium. Most importantly, the batteries themselves become less effective when cycled at repetitive rates. This capacity degradation, often called the *memory effect*, is reversible with reconditioning but could prove to be an annoyance for an EV user who wants a consistant driving range.

SODIUM SULFUR

Considered an emerging technology, sodium sulfur batteries are made up of a molten sodium anode, a molten sulfur cathode, and a solid material sandwiched between that is a sodium ion-conducting electrolyte. These batteries must operate at extremely high temperatures (between 300 and 350 degrees Celsius) and be contained in a well-insulated thermal wrapping. If the battery pack is allowed to fall below these temperatures for extended periods of time, the battery lifespan could be seriously impaired. Because of the extreme reactivity of sodium and sulfur with oxygen and water, care should be taken during replacement or maintenance of the batteries, which only last about half as long as more mature batteries. Despite the need of these batteries to operate in a high temperature environment, automakers continue to explore the use of sodium sulfur because of the potential to dramatically extend vehicle range between recharges. Another promising feature of these batteries is that the raw materials are inexpensive, although a disposal infrastructure would need to be considered before they were widely used in the EV market. So for now, its high operating temperature and limited lifecycle are stalling its use.

NICKEL IRON

Nickel iron batteries are a relatively new addition to the vehicle battery field. They use metallic iron on the negative electrode, nickel oxide on the positive electrode, and a potassium hydroxide solution spiked with lithium hydroxide as an electrolyte. Since they generate significant amounts of hydrogen, adequate ventilation procedures should be used during indoor charging. The batteries tend to offer better performance than lead-acid in several categories. They have a longer life span, can withstand more over-charging and short-circuting, can operate in a wider temperature range of -10 to 45 degrees Celsius, and offer slightly higher power density. However, the batteries need

Type	Advantage	Disadvantage
Lead acid	Low Price Based on established technology Abundant raw materials Adequate cycle life Maintenance free, sealed versions available Recycling system	Low specific energy and energy density increases weight and volume and decreases range Power decreases as battery discharges Decreased capacity at low atmospheric temperatures Limited possibility of fast charging
Nickel cadmium	High Cycle life High power even after partial discharge Very good performance at low atmospheric temperatures Fast charge technology developed	Expensive Cadmium is rare and highly toxic Poor chargeability at high atmospheric temperatures Charging memory effect can systematically reduce capacity No recycling system
Nickel metal hydrides	Very high specific energy and energy density decreases weight and volume and increases range High power Good performance at low atmospheric temperatures	Expensive Use of rare metals in some instances Very sensitive to high temperatures (thermal management required) No recycling system
Sodium sulfur, sodium nickel chloride	Very high specific energy and energy density decrease weight and volume and increases range High power Abundant raw materials	Expensive Premature failures and self-discharges Must maintain permanently high temperatures High internal resistance reduces specific power Safety issues regarding chemical composition and high temperatures No recycling system
Metal air	High specific energy decreases weight and increases range Consistently high power Fast mechanical recharge High safety potential Manufacturing ease Abundant and low-cost raw materials	Currently high cost Hydrogen build up with overcharge Poor performance at low temperatures Low cell efficiency Requires carbon dioxide scrubber
Ambient temperature lithium	High specific energy decreases weight and increases range High power Abundant lithium supply Low maintenance	Carbon version expensive Solid version has low power at room temperature Questionable safety of charging Limitations in quick charging No recycling system

Table 9-1   Advantages and Disadvantages of Current EV Batteries.   Source:
General Accounting Office

State	Total new cars and light trucks registered in 1990	1998	1999	2000	2001	2002	2003
California	1,221,800	24,400	24,400	24,400	61,100	61,100	122,200
Delaware	47,100	900	900	900	2,400	2,400	4,700
Maine	44,300	900	900	900	2,200	2,200	4,400
Maryland	290,00	5,100	5,100	5,100	12,800	2,800	5,600
Massachusetts	255,800	5,100	5,100	5,100	12,800	12,800	25,600
New Hampshire	55,800	1,100	1,100	1,100	2,800	2,800	5,6500
New Jersey	405,600	8,100	8,100	8,100	20,300	20,300	40,600
New York	644,700	12,900	12,900	12,900	32,200	32,200	64,500
Pennsylvania	490,400	9,800	9,800	9,800	24,500	24,500	49,000
Rhode Island	27,400	700	700	700	1,200	1,200	2,400
Vermont	24,300	500	500	500	1,200	1,200	2,400
Washington, D.C.	19,200	400	400	400	1,000	1,000	1,900
Total	3,536,400	70,600	70,600	70,600	176,900	176,900	353,600

Table 9-2   Projected EV Sales. Source: The Electric Power Research Institute

frequent watering, substantial ventilation when recharging, and are more expensive. They also self-discharge at room temperature between 20 and 40 percent per month. Still, its ability to withstand both physical and eletrical abuse is encouraging enough for research to continue.

ENVIRONMENTAL

The most significant advantage of electric vehicles is the immediate environmental benefit from reduced mobile source emissions. Hydrocarbon and carbon monoxide emissions from EVs are typically 10 to 20 times lower than those from ICE vehicles. Other ozone precursors such as nitrogen oxide are also much lower than any other available fuel on the market today. As in the case of California, if EVs are charged by electric utilities that are fueled with hydropower, nuclear power, or other renewable resource, the net effect on the environment is almost no nitrogen oxides, sulfur dioxides, or carbon dioxide — pollutants that can exacerbate ground-level ozone levels and contribute to global warming. However, outside of California, emissions may increase as a result of EVs because some of the energy imported to California comes from Southwest United States coal-fired power plants.[11] The growing use of natural gas in power generation may help mitigate this effect, along with tightening controls on all powerplant emissions.

Using EVs has also been touted as a way to control greenhouse gas emissions. Government estimates suggest that based on the anticipated power mix in the year 2000, an EV would produce 28 percent less carbon dioxide than a comparable gasoline-powered vehicle.

Another environmental consideration is battery disposal. Expanding EV use would also dramatically increase the use of hazardous materials such as heavy metals that require special handling when discarded.

CONCLUSION

Once technologically mature, EVs could easily supplant other alternative fuel-powered vehicles. But at present, there are too many technical and economic barriers for motorists to consider using EVs unless in a fleet environment where federal and state tax incentives are present. Several types of batteries and fuel cells are under serious investigation that may allow EVs to compete or even supplant gasoline-powered vehicles, but none show immediate commercial potential. Also, consumer experience with EVs remains limited. Most demonstrations have dealt with a committed niche market of short-distance drivers already committed to public transportation. Additionally, in the case of California and Massachusetts, significant incentives were provided, such as free AAA membership and a cellular phone for emergencies. EV supporters note that EVs are specifically targeted to a small niche for urban commuters not dependent on a vehicle for varied travel. Driver satisfaction surveys of EVs are very encouraging, although not widespread enough to calm the fears of the auto industry.

References

1. Himy, Albert. *Overview of Various Types of Batteries for Electric Vehicle and Solar Applications*, NESEA S/EV 94 Conference, October 1994.
2. EIA. *Alternatives to Traditional Transportation Fuels: An Overview*, June 1994.
3. University of Wisconsin. *Alternative Fuels Refueling Station Infrastructure Study*, April 1, 1994.
4. Ochs, Dr. M. *Drive Train Products for Environmental Vehicles*, SAE/Environmental Vehicles Conference 95, January 1995.
5. CalFuels Plan. *Developing an Infrastructure Plan for Alternative Fuel Vehicles,* California Energy Commission, September 1994.
6. Murphy-Lessor. *What is the Consumer Demand for Electric Vehicles?* 12th International Electric Vehicle Symposium, December 1994.
7. Haslund, Christina. *Evaluating EV Charging Infrastructure*, 12th International Electric Vehicle Symposium, December 1994.
8. Boivin, Claude. Hydro-Quebec. *Hybrid vs. All Electric: The Test of Performance*, 12th International Electric Vehicle Symposium, December 1994.

9. Hammel, Carol. National Renewable Energy Laboratory. *Identification of Safety Concerns for EVs*, 12th International Electric Vehicle Symposium.

10. Chan, Kwai-Cheung. *Electric Vehicles: Likely Consequences of U.S. and Other Nations' Programs and Policies*, U.S. General Accounting Office, December 1994.

14. Kennedy, John. *Load Management and Power Quality Issues,* 12th International Electric Vehicle Symposium.

C H A P T E R   T E N

HYDROGEN

With environmental concerns expected to stay in the forefront of public policy, a growing amount of research is being spent on fuel sources that are ecologically friendly. While potential fuel sources such has liquid hydrogen and solar power are not part of the energy mix at this time, automakers are expanding research and development efforts in this area. Solar power, which by itself cannot be considered a transportation fuel, is a potential electricity transportation source and as such is discussed in the previous chapter. Hydrogen, however, can be used in both a traditional internal combustion engine (with heavy modifications) or in a fuel cell which could power an electric vehicle. For this reason, hydrogen continues to show promise to policymakers and scientific types because it is the simplest, naturally-occurring element that can be produced from numerous materials, including natural gas, methanol, coal, biomass, and water. It was first used to take the astronauts to the moon and is still used to fuel space missions, according to the records of the National Renewable Energy Laboratory. Because of its environmental benefits, hydrogen is being evaluated for future use in ground transportation. With zero emission vehicles required in California and possibly other states, hydrogen alone or in a fuel cell could meet the regulations. However, there are still too many economic and technical barriers that still need to be addressed, especially specific limitations in engine performance, storage, delivery infrastructure, and production cost. Nevertheless, several countries, including Germany and Japan, already have produced hydrogen-powered demonstration vehicles. But in the United States, policymakers and hydrogen supporters don't expect to see the fuel used in commercial applications for 20 years or more.

FUEL CHARACTERISTICS

From a chemical standpoint, hydrogen (H_2) is the simplest and most naturally-occurring element on the planet. It is a colorless, odorless, and tasteless gas at room temperature, but it can be liquefied for fuel purposes if necessary. When used as a fuel, hydrogen typically contains small amounts of oxygen and other materials, depending on its feedstock source.

Generally, hydrogen can be produced two ways: directly from any hydrocarbon fuel, such as oil, coal, alcohol, biomass, and natural gas, or indirectly from electricity. To directly produce hydrogen, a producer would create a synthetic gas from steam, reforming natural gas or other feedstock such as biomass or coal. Synthetic gas can also be formed through a partial oxidation process. The most environmentally benign method is by decomposing water through electrolysis — application of an electrical current through water so that the hydrogen separates from the oxygen. Heat, chemicals, and even bacteria can be used to split water into its basic elements. However, government researchers predict that future hydrogen production for the transportation market will likely come from nonelectric, direct production paths such as biomass gasification because these methods require less energy and have potentially less capital investment.

STORAGE AND DISTRIBUTION

Currently, a fuel distribution system for hydrogen does not exist although if the fuel becomes more popular, it could theoretically be transported via a pipeline system similar to natural gas. Those that use hydrogen for demonstration projects now rely on canisters and tanker trucks for storage purposes.

Whether used as a liquid in an internal combustion engine or as a fuel cell source, on-board fuel storage is a major stumbling block that will require advanced technology to perfect. As in the case of natural gas, part of the future success of hydrogen will lie in how similar hydrogen fuel storage requirements are to gasoline. Most demonstration vehicles only have one-third the fuel range of a conventional gasoline-powered vehicle.

There are a number of storage options available for hydrogen-powered vehicles, all of which affect overall performance. These include compressed hydrogen, liquid hydrogen, and the chemical bonding between hydrogen and a storage material, such as in gas-solid adsorption and metal hydrides. Compressed hydrogen gas has a low density, which demands a high weight-to-volume ratio. Recent lightweight composite fuel tanks now being considered for natural gas vehicles may help meet motorist expectations for fuel range in the future. Liquefied hydrogen stored in refrigerated tanks provides a longer range, but storage tanks may have limited use because of the fuel properties of hydrogen which result in a continuous boil off. Gas-solid adsorption is an experimental technique that offers higher fuel density than compressed gas. Hydrogen storage using activated carbon is now being researched although the weight of the fuel tanks may pose a problem with fuel range. Another chemical storage system being pursued is the use of metal hydrides. This system appears to offer high storage density and the maximum safety of any on-board storage option. But a big drawback is that low-cost, high-capacity hydrides need very high temperatures to liberate the stored hydrogen. Other hydrides can be used that operate at lower temperatures, but they are more expensive and less efficient, according to research by the National Hydrogen Association.

One option being examined by policymakers would be to offer a hydrogen transition fuel in the marketplace that could take advantage of existing natural gas pipelines. Hythane vehicles would run on a fuel mix of 15 percent hydrogen and 85 percent natural gas, and little modification to an existing NGV would be necessary. If fuel cell vehicles enter the marketplace, the natural gas/hydrogen blends could be converted to hydrogen at the refueling station. Under this scenario, most of the refueling and distribution consideration now in place for natural gas vehicles would also apply to hythane vehicles.

SAFETY CONSIDERATIONS

Anyone who has looked at the famous old film footage of the Hindenburg explosion knows that hydrogen gas, while nontoxic, is very flammable. Since it is still an experimental alternative fuel, safety considerations at refueling sites and throughout the refueling network have yet to be addressed. But as a general rule, automakers looking at hydrogen as a transportation fuel say it should be handled in a similar way to natural gas, which is also very flammable.

At the same time, hydrogen has a minimal impact on land, surface water, and the ocean. Given its capacity to rapidly dissipate into the atmosphere, hydrogen leaks or spills have zero or a negligible impact on underground water, rivers, and lakes.

ECONOMICS

Hydrogen produced from natural gas is the least expensive production method available today, and yet fuel costs are still twice that of gasoline. However, environmental pressures such as global warming and low-level ozone pollutants or a sudden oil shortage could narrow the gap between the two fuels.

In the future, widespread use of hydrogen is expected to develop primarily through the modification of ICEs and through fuel cells, which are devices that change chemical energy directly into electrical energy. Fuel cells have important advantages over ICEs in that they are zero emission technology, but, from an economic standpoint, fuel cell technology cannot compete with pure hydrogen or hydrogen-natural gas blends.

Both technologies are more energy efficient than spark-ignition vehicles using conventional fuels. According to scientists at the Lawrence Livermore Laboratory, in a hybrid mode where hydrogen is burned in an ICE at a steady state, and where peak power was provided by an energy storage device such as a battery or flywheel, efficiencies of more than 45 percent over gasoline have been observed. Similarly, the electrochemical conversion of hydrogen for power through fuel cells offers the same, and in some cases even higher, energy efficiency than a hydrogen-powered ICE does, but with zero or near-zero emissions.

Those who support hydrogen fuels say the direct cost savings of such increased efficiencies would ensure rapid payback of any investment the private sector or the federal government made to commercialize its use.

ENVIRONMENTAL

Arguments can be made about the role of hydrogen in reducing foreign oil dependence and improving energy efficiency, but the prime motivation behind developing hydrogen transportation fuels is environmental. Hydrogen from a scientific standpoint is almost an ideal fuel because when burned in an ICE it produces significantly less emissions than gasoline. These kinds of hydrogen-powered vehicles emit mainly water vapor. No hydrocarbons or sulfur compounds are produced and only a limited amount of nitrogen oxides. When hydrogen is used in a fuel cell, the benefits become even more dramatic. Only water and heat are produced. For this reason, in the eyes of California policymakers, the expanded use of hydrogen is an idea worth exploring. By introducing fuel-cell powered or battery powered vehicles, there is an enormous potential to eliminate a wide range of pollutants that are ozone precursors and global warming aggravators. These pollutants include reactive organic gases, carbon monoxide, methanol, carbon dioxide, sulfur oxides, and particulate matter.

References

1. Hydrogen Fuel Fact Sheet: National Renewable Energy Laboratory.
2. Testimony on H.R. 655: The Hydrogen Future Act of 1995, Full Committee Hearing, February 1, 1995.
3. The National Hydrogen Association: Hydrogen and Fuel Cells: The Perfect Drivers for Trnsportation.

CHAPTER ELEVEN

FLEET CONSIDERATIONS

W ith energy and environmental laws now taking effect nationwide, federal, state, municipal, and some of the larger private fleets
must add or retrofit vehicles to become cleaner burning in
order to comply with strict tailpipe emission requirements. Before the advent of alternative fuels, a fleet manager's biggest concerns were resale value, purchase price,
and load requirements. But AFVs add another set of variables to the equation, including local alternative fuel supply, miles traveled per day, operation and maintenance,

local climate and terrain, emerging vehicle technology like enhanced battery storage or CNG cylinder increases, and insurance.

OEM vehicles typically emit lower pollutants and last longer than retrofitted vehicles because OEM vehicles can be optimized to take advantage of an alternative fuel's specific qualities. However, with widespread introduction of AFVs still evolving, a fleet manager wishing to purchase a light-duty AFV through an automaker is constrained by limited availability and product lines. Part of the reason the selection is limited is due to the small portion of the market devoted to AFVs. Of the 180 million light-duty vehicles, with a gross weight of 8,500 pounds or less, registered in 1992, only 250,000 were AFVs.[1]

Alternative fuel vehicles can operate on one fuel, as a dedicated vehicle, two fuels stored separately as a bifuel vehicle, or a mix of two fuels as a flexible-fuel or hybrid vehicle. Automakers are actively plowing research money into improving the performance and price of AFVs so they can eventually compete with conventional vehicles. Issues that continue to be addressed include vehicle cost, durability, service support, range, performance, and safety.

This means that in the immediate future, many operators are choosing to convert existing vehicles, and the most popular and least expensive fuel conversion option is propane or CNG because of fuel cost, state and federal incentives, and availability.

Fuels	Make	Model	Classification	Power Train	Fuel Capacity
M85 and gasoline	Chrysler	Dodge Intrepid	Mid-size car	3.3 Liter/V-6	18.0 gallons
M85 and gasoline	Ford	Taurus	10B Full-size car	3.0 Liter	20.4 gallons
CNG	Chrysler	Ram Van/Wagon	Full-size van	5.2 Liter/V-8	14.4 or 15.7 GGE
CNG	Chrysler	Ram Pickup	Full-size	5.2 Liter/V-8	17.1 GGE at 3,000 psi
CNG	Chrysler	Dodge Caravan/	Minivan	3.3 Liter/V-6	8.2 GGE at 3,000 psi
CNG	Chrysler	Dodge Dakota	Mid-size truck	5.2 Liter/V-8	14.6 GGE
CNG and gasoline	Ford	F-Series, Super Cab 4x4 F150/250	Full-size truck	4.9 Liter/ Inline-6	As much as 12.1 GGE of CNG plus 18.2 gallons gasoline
CNG and gasoline	Ford	Econoline E250 HD, E350	Full-size van	4.9Liter/ Inline-6	Up to 12.1 GGE of CNG plus 35 gallons gasoline
LPG	Ford	F600/F700 Medium-duty Chassis Cab Truck (21,000- 35,000 GVWR)	Medium-duty truck	7.0 Liter, 218 HP	Depends on LPG tank installed by aftermarket converter
Electricity	Chrysler	Dodge/Plymouth	Minivan	70 HP DC (Max), 35 HP DC (continuous)	

Table 11-1   Major U.S. Automakers' AFV Production Plans MY 1995. Source: U.S. Alternative Fuels Hotline

CONVERSION ISSUES

TYPE

With automakers still gearing up to meet AFV demand, a fleet manager often must consider converting an existing or newly-purchased vehicle to operate on an alternative fuel, either alone or in conjunction with conventional gasoline or diesel. There are a number of conversion systems, sometimes called *after-market* fuel delivery systems, on the market. Most systems are similar in design and operation, but it is important to ensure the

Photo 10-1   Dodge Compressed Natural Gas (CNG) Fuel Cylinder

system is installed correctly to comply with clean air laws. Also, a kit should only be installed on vehicles recommended by the kit manufacturer and/or the automaker. It's important to note that many states have enacted tampering laws regulating vehicle conversions. Before converting a vehicle, check with the state air pollution control agency to find out what conversion restrictions apply in a particular area.

A system designed to run on either an alternative fuel or conventional gasoline is called a *bifuel* system. This kind of setup can be helpful when a driver does not have access to an alternative fuel refueling station. However, the motorist can only use one fuel at a time.

Both bifuel and dual-fuel vehicles have two separate fuel tanks. But dual-fueled engines can run on a combination of alternative and conventional fuels. Dual-fuel systems are used mainly in heavy-duty or diesel engines; bifueled systems can most often be found in light-duty passenger cars and trucks. Dedicated conversion systems are designed to run on only one fuel, and as a result generally have the lowest tailpipe emissions because they are optimized for a specific fuel type.[2]

Open-loop conversion systems are found on older vehicles that do not have computerized fuel injection and are slowly leaving the marketplace as cars are retired or reconfigured with more sophisticated fuel control systems that maximize engine performance. These new closed-loop systems provide optimum emissions performance by a computerized adjustment of the fuel/air mixture to the engine. Both open- and closed-loop systems can be found on bifuel, dualfuel, and dedicated engines.

WARRANTY

Converting a gasoline- or diesel-powered vehicle to run on an alternative fuel normally does not cancel or void an automaker's warranty. However, an engine failure caused by the conversion system is usually not covered by the OEM. When looking for a conversion company, make sure the kit manufacturer warranties its parts. Also, be aware some automakers have arrangements with outside conversion companies that complete the installation of the fuel system after a vehicle is built by the OEM. In these instances, the conversion company is certified by the automaker as a qualified vehicle modifier to convert specific vehicle types and models. A big benefit for a fleet manager is that these conversions allow a car dealer to offer a similar warranty plan for both converted and OEM vehicles.

Operation and Maintenance

The operation and maintenance of an AFV depends on a variety of factors, including type of fuel; whether the vehicle is dedicated, dual-fueled, or flexible fueled; and whether the vehicle came from the auto manufacturer directly or was converted by an outside vendor. There is laboratory evidence that the use of cleaner-burning fuels may help improve an engine's life cycle by two to three times, but AFVs don't yet have the track record to prove this claim yet.

The following is a brief discussion of special concerns associated with the operation of nontraditional transportation fuels. Reformulated and oxygenated fuels are not included in this discussion since these two fuel types are compatible with conventional gasoline-powered engines.

ALCOHOL FUELS

Perhaps the biggest selling point for alcohol-fueled vehicles is that to the motorist, both ethanol and methanol for the most part offer the same performance and driveability as gasoline vehicles. As with a conventional gasoline-powered vehicle, a dedicated or flexible fuel vehicle is calibrated to provide maximum engine performance. However, dedicated vehicles may offer slightly improved performance over their flexible fuel counterparts since FFVs must be calibrated to adapt to any combination of alcohol or gasoline. FFVs must accommodate the lowest octane fuel available so the compression ratio is not increased to yield the full engine efficiency benefits of ethanol or methanol.

In the case of a dedicated vehicle, neat alcohol also offers a slightly higher power output than gasoline vehicles because of greater fuel efficiency and higher compression ratios. Both ethanol and methanol fuels are best suited for spark-ignition engines because of their high octane value and relatively low cetane value. Methanol and ethanol can be used in compression-ignition engines designed for diesel fuel, provided the fuel or the engine has been modified through an ignition improver.

Since there are relatively few alcohol-powered cars on the road today, it is difficult to track the repair record of these vehicles. Due to the corrosive nature of alcohol fuels, dedicated and flexible fuel vehicles require a special engine oil and fuel system that are detailed by the manufacturer.

NATURAL GAS

Most gasoline or diesel-powered vehicles can be converted to operate on CNG or LNG exclusively or in combination with traditional petroleum fuels. The physical and chemical properties of natural gas provide good engine performance with power losses of less than 10 percent in the case of some converted vehicles. This is because natural gas displaces more air in the combustion chamber than gasoline. Gasoline is not entirely vaporized when introduced into the fuel chamber and therefore takes up less area than a gaseous fuel.

OEM produced CNG vehicles tend to be cleaner burning but have no significant other benefits over a converted vehicle. CNG conversion kits range from $2,500 to $4,000; the cost will vary depending on the number of tanks used, the size of the tanks, and the material used to make the tanks.

In CNG vehicles, gas is stored at high pressures of 2,400 to 3,600 psi in one or more cylinders usually located in the rear or trunk of a vehicle. The filling valve is placed near the tank. When the gaseous fuel leaves the cylinder tank, it travels through high-pressure fuel lines into one or more pressure regulators where the CNG is reduced to almost atmospheric pressure. A valve opens to allow the gas to enter the combustion chamber, and it is then ignited to create power to propel the vehicle.

NGVs require special maintenance not found with conventional fuels due to the high-pressure nature of the fuel system. CNG cylinders, for example, must be periodically recertified to maintain compliance with Department of Transportation standards. NGV operators using a dual-fuel system should also run the vehicle on gasoline for a few miles each week to prevent drying and cracking of the gasoline system components.

Also, it is important to change oil regularly on an NGV.

Oil degrades at about the same rate with gaseous fuels as it does with gasoline, so extending the oil change interval could damage the engine.

LIQUEFIED PETROLEUM GAS

Although automakers at one time made LPG-powered vehicles at the factory, no manufacturer is currently producing a light-duty passenger vehicle for sale in the United States. As a result, fleet managers have had to rely on aftermarket conversions. These conversions are sophisticated enough to allow for the efficient use of the fuel with nearly no change in engine performance. To convert a light-duty vehicle to dedicated LPG use, a high-strength fuel tank is added along with a fuel-lock filter, shutoff valve, vaporizer/regulator, air-fuel mixer, and microprocessor for air/fuel ratio control. LPG conversion kits cost between $2,000 and $2,500, significantly less than a CNG conversion which requires more sophisticated cylinders to accommodate high pressure gas.

LPG maintenance is similar to CNG. Even though oil used in an LPG engine may appear clean in visual inspections, it's important to change the oil according to manufacturer recommendations since oil components break down under engine heat, causing lubricity breakdowns. Additionally, spark plugs should follow regular replacement schedules, even though fouling is not an issue with clean burning fuels, because spark plugs still wear out with firing.

ELECTRIC VEHICLES

EVs are unique among other alternative fuel vehicles in that they do not require an internal combustion engine. Instead, the power system includes a motor to convert electrical energy to mechanical energy to power the vehicle, battery pack, controller to regulate electric flow from the battery pack to the motor, converter, watering cart, and battery charger. Hybrid vehicles, which use one or more power sources to extend range, may have a battery and a small internal combustion engine with a small fuel tank, or a fuel cell and a battery, or a flywheel and a battery.[3]

An often-touted advantage of electric vehicles over their gasoline-powered cousins is lower maintenance. Maintaining an EV is estimated to take half the time of an internal combustion engine because there are fewer mechanical parts. While there may be no oil to change, there is still preventive maintenance that must be performed. The battery pack must be recharged daily, for example. Current battery types also require water every three weeks, but battery types under consideration may do away with this chore. Battery packs also must be replaced every few years, depending on battery type and need to be disposed of in an appropriate landfill site. Care should also be taken to watch for acid leakage and to avoid electric shock.

TRAINING

Vehicle equipment installation training has traditionally been available through retrofit kit manufacturers and distributors. As emissions requirements become more precise, it is important that installers and service technicians are properly trained. To that end, the government is now in the process of spearheading a more uniform AFV maintenance certification process so there will be a sufficient number of readily available, skilled AFV technicians to perform conversions and to maintain and repair vehicles and refueling systems.

Section 411 of the Energy Policy Act directs the Secretary of Energy to establish a national program to certify alternative fuel training programs for automotive technicians. Referred to as "CHAMP," the voluntary certification program was to be launched in the fall of 1995 through a recent 5-year cost-shared cooperative agreement with the National Automotive Technician Education Foundation. Working with the alternative fuels industry and DOE, the foundation will establish national uniform standards and manage the certification process. It will also perform evaluations and make recommendations for certification. Based on the foundation's recommendations, the nationally recognized, independent private sector automotive certifying organization, the National Institute for Automotive Service Excellence (ASE), will confer the certifications under the program. The foundation will also provide program information to training providers through written materials and regional workshops.

RESALE

One of the biggest reasons fleet managers have been hesitant to purchase AFVs is that the resale value of these vehicles is unclear. Nearly all AFV purchases are motivated by government intervention, but fleet managers still must consider a number of factors, including mileage, emission level, safety, maintenance, and operating expense, that would be considered by a potential buyer. It is expected that demand for used AVFs will be from other fleets, mainly utilities and government operators that must comply with regulatory requirements. In the case of a converted vehicle, it's best to stick with a factory or OEM-approved conversion company so the vehicle will remain under warranty and appeal to a broad market.

HEAVY-DUTY APPLICATIONS

Nearly all alternative fuels have fuel characteristics that fit with the design parameters of spark-ignition, light-duty vehicles. In heavy-duty compression-ignition engines that typically run on diesel, higher pressures and temperatures allow for greater fuel efficiency, which in turn promotes longer engine life. Since the heavy compression-ignition engine can last longer between major overhauls, diesel engines are favored by commercial transport, which is willing to trade the higher upfront cost of the vehicle for longer vehicle miles travelled. A major obstacle for heavy-duty fleet operators, therefore, is how to maximize fuel performance using alternative fuels mainly suited for light-duty applications in spark-ignition engines.

Responding to fleet customer's needs to comply with clean air legislation, nearly all of the nation's heavy-duty engine manufacturers are offering alternative-powered compression-ignition engines. Engine manufacturers such as Hercules, Cummins, Caterpillar, and Detroit Diesel Corporation have developed and are demonstrating alternative-fueled engines in a variety of applications, including school buses, urban transit buses, and delivery vehicles. Chassis manufacturers such as Crane Carrier have formed alliances with these engine makers who are also using Thomas Built and Blue Bird school buses and Flxible and Bus Industries of America transit buses.

The largest target market for diesel fuel alternatives is transit buses. There are three major reasons for this. Unlike heavy-duty trucks which often run on gasoline, urban buses have been powered almost exclusively by diesel engines for the past 50 years. Transit authorities, unlike the deregulated trucking industry, often have as much as 80 percent of their operating expenses covered by the state, municipal, and federal governments. This gives them the flexibility to experiment with various alternative fuels. Heavy-duty truck operators, on the other hand, work in an industry in which thin margins, competition, and overcapacity of equipment do not allow much room for any alternative fuel unless it is cost-competitive with gasoline. Under the Clean Air Act Amendments of 1990, urban buses have been singled out from inter-city coaches and heavy-duty trucks to meet stringent NO_x and particulate standards .

Spurred by these environmental considerations, about 10 percent of the nation's transit buses and vans currently operate on alternate power. As clean fuel fleet requirements tighten, that number is expected to double before the end of the decade. Of the thousands of buses, vans, other demand-response vehicles, and electric trolleybuses now running on alternative fuel technologies, 40 percent operate on diesel fuel with particulate traps, 28 percent on electricity, and 8 percent run on propane. The remainder are powered with CNG, 6 percent; methanol, 5 percent; liquid natural gas, ethanol and battery electric, 1 percent each; or by dual-fueled engines such as those that use diesel and ethanol or gasoline and CNG, 6 percent.

OFF-ROAD MARKET OPPORTUNITIES

LOCOMOTIVES

The railroad industry already operates its own kind of hybrid engine. Diesel/electric locomotives have been a mainstay of the industry for years as a way to lower fuel costs. It is unlikely that diesel-powered engines will be replaced by alternative fuels for a quite some time. Nevertheless, some rail executives consider natural gas to be a good replacement fuel because of price. Burlington Northern, for example, is working with two 3,000-horsepower (hp) prototype locomotives that run on LNG. The fuel is stored in 3,000- to 5,000-gallon saddle tanks under the train's frame in the form of refrigerated liquid methane. The demonstration is being run with the help of Air Products, which supplied the fuel and helped Burlington convert a General Motors Electro-Motive 970s diesel engine for dual-fuel use. The locomotive requires diesel at idle and as a pilot fuel. At all other times the engine runs on liquefied methane. This particular conversion gives the railroad full-rated horsepower, unlike straight natural gas, which when used in a transportation application whether it be a heavy-duty truck, forklift or train, loses horsepower. Other rail engine manufacturers are studying the possibilities of alternative-fueled locomotives.

Caterpillar, in a joint venture with Morrison Knudson, has designed a liquefied natural gas switcher locomotive rated at 1,200 hp for traction instead of the usual 2,000. Also, both General Motors' and General Electrics' transportation system divisions have on the drawing board a dedicated LNG locomotive system.

Also, if engineers can find a way to boost fuel injection pressure, the need for a pilot light could be eliminated and fuel costs decreased enough to make natural gas an attractive option. Of course, even the best ideas can be thwarted by economics, and in the case of the beleaguered railroad industry, analysts say this technology is at least five to 10 years away from the station. The most economical situation for a natural gas locomotive would be for long distance freight runs. Using natural-gas-powered trains in passenger services may take some time because of public concerns of sitting near a natural gas tank.

FORKLIFTS, MINE VEHICLES, AND FARM EQUIPMENT

Off-road diesel engines represent a small but significant market to diesel producers. According to the Society of Automotive Engineers, more than 315,000 engines in 1989 were supplied to this market, with construction/mining representing 50 percent of off-road engines, 43 percent agricultural, 1 percent stationary source generators, and 6 percent marine.

At this writing, EPA and its state counterparts were expected to expand their nets of regulation to include virtually any kind of petroleum-fueled engine — construction equipment, farm machinery, even lawnmowers are not immune to scrutiny. However, government air pollution officials are meeting resistance from individual farmers, off-road engine manufacturers, and major oil companies about proposed emission standards for tractors and other heavy equipment powered by diesel engines.

EPA estimates the clean-air requirement would raise the cost of a typical diesel farm engine by $110. But manufacturers believe the added cost will be at least twice that, and in some cases as much as 10 to 20 times more. Farmers, are concerned that tractors designed to run on natural gas or alcohol will be less efficient and drive up fuel costs. The American Farm Bureau Federation (AFBF) says the cleaner engines will consume 3 percent to 5 percent more fuel. That would cost barley and wheat farmers an extra 26 cents per acre, corn growers an extra 38 cents per acre, cotton farmers $1.09 an acre more, and rice farmers an additional $2.12 an acre, AFBF says.

Until now, there has not been much of a market for diesel fuel alternatives in outdoor construction equipment and farm machinery because EPA has long exempted them from clean air regulations. But with many state and local regulators questioning whether the proposed diesel standards are stringent enough, manufacturers are eyeing natural gas, alcohol, and electrically-powered engines for California and other niche markets where a diesel fuel alternative may be mandated.

Farm machinery accounts for at least 10 percent of the nitrogen oxide emissions in California's San Joaquin Valley, one of 25 polluted areas studied by the EPA. Bulldozers and other construction equipment are considered a major polluter in many big cities. In the Northeast, off-road engines account for 15 percent of all nitrogen oxide emissions, according to state regulators.

The Equipment Manufacturers Institute (EMI) argues that EPA has overestimated the role of off-road equipment and air pollution by assuming that sales of tractors and construction equipment will grow 2 percent a year. Sales actually have been declining for the past two decades, according to EMI.

EPA also has underestimated the cost of redesigning engines, the group contends. It estimates customers will pay $270 to $2,900 more, depending on the type of equipment. Compact Loader Owatonna's Mustang Manufacturing Co. estimates that the compact loaders it sells to farmers and building contractors will cost an extra $1,156, or about 5 percent more. Modifications to the engine design alone will cost $500 each, according to the company. Mining companies have joined with farmers in arguing that their equipment isn't responsible for enough pollution to justify the emission controls. Mining equipment will cost $3,000 to $30,000 more per engine, according to the American Mining Congress.

While off-road engine builders may not be happy about EPA's regulatory temperament, they have several programs in the works to comply with encroaching government regulations. Detroit Diesel offers a low emission, electronically controlled diesel engine for underground mining equipment.

For other indoor engine applications such as lift trucks and forklifts, the natural gas industry has taken the lead in research and development, primarily through the Gas Research Institute. Companies that operate lift trucks fueled by propane, gasoline, or diesel fuels must consider OSHA regulations on employee carbon monoxide exposure in addition to other proposed regulations on NO_x and hydrocarbons.

References

1. Ingram, Mike. "Training Programs for Alternative Fuel *Vehicles,*" *Utility Fleet Management,* July/August 1995.
2. Vogelheim, Charlie. "Projecting Residual Values for Alternative Fuel Vehicles," *Automotive Fleet,* April 1994.
3. AFDC Update. "Federal Express Project Ends After Two Years of Data Collection," Spring 1994.

A P P E N D I X   I

LEGISLATION IMPACTING ALTERNATIVE FUELS

THE CLEAN AIR ACT AMENDMENTS OF 1990

The Clean Air Act Amendments (CAAA) of 1990 do not specifically require the use of alternative-fueled vehicles, but they do include strict tailpipe emission standards for passenger cars and light-duty trucks, which are designed to encourage the use of nontraditional fuels. In addition, the CAAA require the use of reformulated gasoline and oxygenated gasolines in certain parts of the country, vehicle fueling emissions controls, new standards for heavy-duty vehicles and urban buses, and several alternative fuel vehicle programs.

Of particular importance to vehicle fleet operators is the CAAA's establishment of the Clean Fuel Fleet Program (CFFP). The CFFP applies to air quality nonattainment areas classified as serious, severe, and extreme for ozone, and moderate and serious areas greater than 12.7 parts per million for carbon monoxide.

Under section 246 of the CAAA, 22 metropolitan areas in 19 states must revise their state implementation plans to include clean-fuel vehicle fleet programs. Beginning in 1998, the program requires that a portion of new vehicles purchased annually by certain fleet owners operating in ozone and carbon monoxide nonattainment areas must be clean-fuel vehicles.

Currently, reformulated gasoline cannot meet the EPA's tough vehicle emission standards. However, the agency expects that air pollution control improvements by both automakers and the oil industry will qualify the use of gasoline before the start of the program. Fleets of 10 or more vehicles must comply except for the following: public rental vehicles, law enforcement and emergency vehicles, nonroad vehicles, vehicles garaged at a personal residence when not in use, and heavy-duty vehicles.

Fleet managers must phase the program in over the course of three years, with the rate determined by vehicle type or weight. For light-duty vehicles and trucks, the phase-in rate increases from 30 percent of new vehicle purchases in 1998 to 50 percent in 1999 and finally 70 percent in the year 2000 and beyond. At least half of a new heavy-duty vehicle fleet must be powered by a clean fuel in 1998 and every year thereafter.

Fleet owners can fulfill their clean fleet requirements in a number of ways. They can purchase clean-fuel vehicles that meet the equivalent of California's low emission standard (LEV). Reformulated gasoline is expected to qualify. Or, fleet owners can use a cleaner fuel such as CNG and earn credits that can be redeemed or sold to other fleet operators.

In addition to establishing the CFFP, the CAAA allowed California its own more stringent regulations and gave other states the option of adopting California's tailpipe emissions standards.

CALIFORNIA'S LOW-EMISSION VEHICLE AND CLEAN FUELS REGULATIONS

Provisions made in the CAAA allowed California to adopt regulations that require auto manufacturers to produce low-emission vehicles. These regulations spurred the development of AFVs. In 1990, the California Air Resources Board adopted the California Low-Emission and Clean Fuels (LEV/CF) program that made automobile emission regulations more stringent than the national standards of the CAAA.

The LEV/CF regulations took effect in model year 1994

and become increasingly stringent through 2003. Relative to today's new vehicle exhaust emissions standards, the LEV/CF regulations lower nonmethane organic gas emission standards for new vehicles by 40 percent in 1998 and by more than 75 percent in 2003. The clean burning characteristics of alternative fuels provide manufacturers with additional methods to meet these standards.

REFORMULATED GASOLINE AND OXY-FUELS

Another portion of the CAAA which impacts, fuel suppliers and consumers is the clean fuel program designed to reduce carbon monoxide and ground level ozone levels.

To reduce ozone levels and toxics, EPA has required since January 1, 1995, that reformulated gasoline (RFG) must be sold in the nation's worst ozone nonattainment areas. These nine metropolitan areas of the country, which represent close to 25 percent of the nation's fuel supply, include Baltimore, Chicago, Hartford, Houston, Los Angeles, Milwaukee, New York, Philadelphia, and San Diego.

Eighty-seven other ozone nonattainment areas may choose to adopt the program on a voluntary basis through a procedure called *opt-in*. Today, more than a dozen states have already decided to opt-in to RFG, although concerns over the fuel's higher price have led some states to pare back their programs.

The agency can delay an opt-in area for three one-year periods if sufficient domestic capacity for RFG is not available. EPA also has the right to waive the requirements of the program for an area if air quality is not severe enough to justify a program.

Because of the extent and nature of air pollution in California, federal regulators are expected to allow tougher state RFG rules to take effect in 1996. California's Clean Air Act requires a 55 percent reduction in VOCs and 15 percent reduction in NO_x by the end of 2000.

When the final RFG specifications are finalized, gasoline refiners will have the option of certifying their own fuel performance standard, provided it meets EPA emission standards. Otherwise, they must follow a prescribed fuel formula. Under the fuel formula, a refiner must sell a gasoline that displays the following characteristics:

■ engine deposit control additives
■ maximum benzene content of 1 percent volume with averaging/credit trading
■ minimum oxygen content of 2 percent by weight with averaging/credit trading
■ no lead content
■ maximum aromatic content of 25 percent volume with averaging/credit trading

With regard to the performance standard, a refiner must be able to prove his fuel achieves a 15 percent reduction, measured on a mass basis, for VOCs and toxics. The standard increases to 25 percent VOCs and toxics reductions by the year 2000 at the discretion of EPA. EPA adopted two models for refiners who wish to meet a performance standard. The simple model involves two fuel parameters, oxygen content and volatility, and can be used for RFG sold in 1995 and 1996.

Starting in 1997, refiners wishing to certify fuels using a performance standard will have to meet the yet-to-be-determined complex model, which will include a number of inputs including auto/oil air quality improvement research program data.

EPA also has a carbon monoxide clean fuel program that requires metropolitan areas that violate federal carbon monoxide standards to sell gasoline containing at least 2.7 percent oxygen by weight during the winter months.

THE ENERGY POLICY ACT OF 1992

TITLE III: ALTERNATIVE FUELS, FEDERAL PROGRAMS

Under this provision, Congress created a combination of mandates, incentives, research and development programs, and public information campaigns designed to encourage the expanded use of nonpetroleum fuels.

Federal Fleet Requirements

The legislation requires the federal government to help establish an AFV market by purchasing 5,000 AFV light-duty cars and trucks in 1993, 7,500 in 1994, and 10,000 in 1995. In 1996, the requirement changes to 25 percent of new vehicle purchases. By 1999, 7 percent of all new federal light duty vehicles must be capable of running on an alternative fuel.

Since there are about 375,000 light duty vehicles in the federal fleet, with about 20 percent annual turnover, the new law should result in about 125,000 to 160,000 AFVs by the end of the decade.

State Fleets

The law requires centrally fueled state government fleets in metropolitan areas of 250,000 or more to buy AFVs on the following schedule: 10 percent of new vehicles in model year 1996, 15 percent in 1997, 25 percent in 1998, 50 percent in 1999, and 75 percent in 2000 and all future years.

PRIVATE AND MUNICIPAL FLEETS

P rivate and municipal fleets of 50 or more vehicles with at least 20 vehicles in one location and in metropolitan areas with a population of 250,000 or more people may be required to phase in light-duty AFVs, starting with 20 percent of new vehicles purchased in model year 1999. Before the program can start, the DOE must decide by December 1996 whether a private fleet program is needed to achieve a goal of replacing 30 percent of imported motor fuels with domestic fuels by the year 2010. Up to 7 million private, centrally fueled fleets could be included in this program, which would dramatically increase demand for alternative fuels, particularly compressed natural gas.

Alternative fuel providers, including public utilities, are required to phase in light-duty AFVs, beginning with 30 percent of their new fleet vehicles beginning model year 1996. The requirement moves up to 50 percent in model year 1997, 70 percent in model year 1998, to 90 percent for model year 1999 and after. Public utilities, including gas, electric, water, and telephone, account for 1.2 million vehicles.

TAX INCENTIVES

U nder the legislation, the incremental cost of AFVs over gasoline and diesel vehicles can be deducted at the following rates: $2,000 for light-duty vehicles, $5,000 for vehicles between 10,000 and 26,000 pounds gross vehicle weight (gvw), and $50,000 for vehicles above 26,000 gvw.

Gasoline marketers and other fueling station owners who install alternative fuel pumping equipment are eligible for up to $100,000 in tax deductions.

OTHER CONSIDERATIONS FOR PETROLEUM MARKETERS

The law allows retailers to sell natural gas at the pump without being subject to regulation as a utility. Also, a low interest loan program by the federal government funds the incremental costs of vehicle conversions and purchases.

THE ALTERNATIVE MOTOR FUELS ACT OF 1988

The Alternative Motor Fuels Act (AMFA) of 1988 sought to encourage the development of alternative fuels through the production of vehicles designed to run on these fuels either in combination with or instead of gasoline.

Starting this year, AMFA allows automakers producing AFVs special credits which count toward corporate average fuel economy (CAFE). The law also requires the demonstration and testing of alternative fuels in a variety of vehicles.

A P P E N D I X   I I

SIDE-BY-SIDE
COMPARISON OF
EPACT AND CAA

Table 1
New Fleet Vehicle Purchase Required by EPACT/CAA

	Clean Air Act		Energy Policy Act			
Year	GVW Less Than 8,500 lb (% of CFVs)	GVW Less Than 26,000 lb (% of CFVs)	Federal (b) (% or Number of AFVs)	State (% of AFVs)	Fuel Provider (% of AFVs)	Municipal/ Private (c) (% of AFVs)
1993			7,500 (a)			
1994			11,250 9a)			
1995			15,000 (a)			
1996			25%	10%	30%	
1997			33%	15%	50%	
1998	30%	50%	50%	25%	70%	
1999	50%	50%	75%	50%	90%	20%
2000	70%	50%	75%	75%	90%	20%
2001	70%	50%	75%	75%	90%	20%
2002	70%	50%	75%	75%	90%	30%
2003	70%	50%	75%	75%	90%	40%
2004	70%	50%	75%	75%	90%	50%
2005	70%	50%	75%	75%	90%	60%
2006	70%	50%	75%	75%	90%	70%

(a) As required by Executive Order No. 12844
(b) Fiscal year for federal fleet acquisitions requirements, model year for all others
(c) May be required if DOE finds these voluntary acquisitions unlikely to be met

Table courtesy U.S. Alternative Fuels Hotline. Source: EPA

APPENDIX III

ALTERNATIVE FUEL

AND

VEHICLE CONTACTS

REFORMULATED
GASOLINE

Terrence Higgins
Technical Director
National Petroleum Refiners Association
Suite 1000
1899 L Street, N.W.
Washington, D.C. 20036
(202) 457-0480

Susan Hahn
Public Affairs
American Petroleum Institute
1220 L Street, N.W.
Washington, D.C. 20005
(202) 682-8118

Doug Durante
Executive Director
Clean Fuels Development Coalition
7315 Wisconsin Avenue
East Tower
Suite 515
Bethesda, MD. 20814
(301) 913-9636

Fred Craft
Executive Director
1330 Connecticut Avenue, N.W.
Suite 300
Washington, D.C. 20036
(202) 296-4200

AUTO/OIL AIR QUALITY IMPROVEMENT RESEARCH PROGRAM

Joe Colucci
General Motors Research Laboratories
30400 Mound Road
Warren, MI 48090

J. Wise
Mobil Research and Development Corporation
P.O. Box 1031
Princeton, NJ 08540

NATURAL GAS

Carol May
NGV Market Development
American Gas Association
1515 Wilson Boulevard
Arlington, VA 22209
(703) 841-8574

Executive Director
Natural Gas Vehicle Coalition
1515 Wilson Boulevard
Suite 1030
Arlington, VA 22209
(703) 527-3022

Ginny Hobbs-Moore
Gas Research Institute
8600 Bryn Mawr Avenue
Chicago, IL 60631
(312) 399-8100

PROPANE (LIQUEFIED PETROLEUM GAS)

Lisa Bontempo
Director
National Propane Gas Association
4301 Fairfax Drive
Suite 340
Arlington, VA 22203
(703) 351-7500

Richard Roldan
Executive Director
Propane Vehicle Council
901 15th Street, N.W.
Suite 470
Washington, D.C. 20005
(202) 371-6262

LP-Gas Clean Fuels Coalition
Bob Myers, Executive Director
2102 Business Center Drive
Suite 130
Irvine, CA 92714

ELECTRIC

Carol Precobb
Electric Power Research Institute
Electric Transportation Information Center
501 14th Street
Suite 210
Oakland, CA 94612
(800) 848-EITC

Kateri Callahan
Executive Director
Electric Transportation Coalition
1050 Thomas Jefferson Drive, N.W.
6th Floor
Washington, D.C. 20007
(202) 298-1935

CALSTART
Micahel Gage, President
3601 Empire Avenue
Burbank, CA 91505
(818) 565-5606

Northeast Sustainable Energy Association
Lisa Petraglia
23 Ames Street
Greenfield, MA 01301
(413) 774-6051

METHANOL

Ray Lewis
President
American Methanol Institute
815 Connecticut Avenue, N.W.
Suite 800
Washington, D.C. 20006
(202) 467-5050

ETHANOL

Eric Vaughn
President
Renewable Fuels Association
One Massachusetts Avenue, N.W.
Suite 820
Washington, D.C. 20001
(202) 289-3835

HYDROGEN

Debbi Smith
Vice President
National Hydrogen Association
1101 Connecticut Avenue, N.W.
Suite 910
Washington, D.C. 20036-4304
(202) 223-5547

BIODIESEL

Bill Holmberg
President
American Biofuels Association
1925 N. Lynn Street
Suite 1050
Arlington, VA 22209
(703) 522-3392

Kenlon Johannes
Executive Director
National Soydiesel Development Board
P.O. Box 104898
Jefferson City, MO 65110
(800) 841-5849

GOVERNMENT

California Air Resources Board
Public Information Office
(916) 322-2990

California Department Of Consumer Affairs
Bureau of Automotive Repair
(800) 952-5210

California Energy Commission
Transportation Technologies and Fuels Office
(916) 654-4634

U.S. Department of Energy
National Alternative Fuels Hotline
P.O. Box 12316
Arlington, VA 22209
(800) 423-1DOE

Jeff Alson
U.S. Environmental Protection Agency
Division of Regulatory Programs and Technology
2565 Plymouth Road
Ann Arbor, MI 48105
(313) 668-4296

U.S. Environmental Protection Agency
401 M Street, SW FOSD (6406J)
Washington, D.C. 20460
Oxy-fuel/RFG Enforcement
Meredith Miller
(202) 233-9031
Al Mannato
(202) 233-9050

GENERAL INFORMATION

California Energy Commission
Transportation Technologies and Fuels Office
1516 Ninth Street, MS-41
Sacramento, CA 95814
(916) 654-4634

National Alternative Fuels Hotline
P.O. Box 12316
Arlington, VA 22209
(800) 423-1DOE

ELECTRIC VEHICLES

Chrysler Corporation
A. Michel Clement
Manager, Alternative Fuel Vehicle Sales and Marketing
27777 Franklin Road
19th Floor
Southfield, MI 48034
(810) 948-3644

Ford Motor Company
John Wallace
23400 Michigan Avenue
Suite 230
Dearborn, MI 48124
(313) 390-5589

General Motors Corporation
Ray Buttacavolie
515 Marin Street
Suite 216
Thousand Oaks, CA 91360
(805) 373-8492

Solectria

Karl Thidemann
Director of Marketing
68 Industrial Way
Wilmington, MA 01887

Specialty Vehicle Manufacturing

Don Duffy
President
9250 Washburn Road
Downey, CA 90241
(310) 904-3434

U.S. Electricar
P.O. Box 6645
Santa Rosa, CA 95406
(707) 525-3227

Methanol/Ethanol Vehicles

Chrysler Corporation
A. Michel Clement
Manager, Alternative Fuel Vehicle Sales and Marketing
27777 Franklin Road
19th Floor
Southfield, MI 48034
(810) 948-3644

Ford Division
Lou Ulrich
Fleet Sales Department
2099 South State College Boulevard
Suite 600
Anaheim, CA 92816
(714) 939-3562

Ford Division
Bill Boultes
Fleet Sales Department
P.O. Box 9048
Pleasanton, CA 94566
(510) 463-5791

General Motors Corporation
Gerald J. Barnes
Manager
Automotive Emissions Control
3044 West Grand Boulevard
Detroit, MI 48202
(313) 556-7723

NATURAL GAS VEHICLES

Chrysler Corporation
A. Michel Clement
Manager, Alternative Fuel Vehicle Sales and Marketing
27777 Franklin Road
19th Floor
Southfield, MI 48034
(810) 948-3644

Ford Division
Lou Ulrich
Fleet Sales Department
2099 South State College Boulevard
Suite 600
Anaheim, CA 92816
(714) 939-3562

Ford Division
Bill Boultes
Fleet Sales Department
P.O. Box 9048
Pleasanton, CA 94566
(510) 463-5791

General Motors Corporation
Gerald J. Barnes
Manager
Automotive Emissions Control
3044 West Grand Boulevard
Detroit, MI 48202
(313) 556-7723

LPG VEHICLES

Propane Vehicle Council
Bob Myers
2102 Business Center Drive
Suite 130
Irvine, CA 92714
(714) 253-5757

A P P E N D I X   I V

FUEL PROPERTIES

Table 1 Properties of Fuels

Property	Gasoline	No. 2 Diesel Fuel	Methanol	Ethanol	MTBE	Propane	CNG (Methane)	Hydrogen
Chemical Formula	C_4 to C_{12}	C_3 to C_{25}	CH_3OH	C_2H_5OH	$(CH_3)_3COCH_3$	C_3H_8	CH_4	H_2
Molecular Weight	100-105[a]	200 (approx.)	32.04	46.07	88.15	44.1	16.04	2.02[x]
Composition, Weight %								
Carbon	85-88[b]	84-87	37.5	52.2	68.1	82	75	0
Hydrogen	12-15[b]	33-16	12.6	13.1	13.7	18	25	100
Oxygen	0	0	49.9	34.7	18.2	–	–	0
Specific Gravity, 60°F/60°F	0.72-0.78[b]	0.81-0.89[d]	0.796[c]	0.796[c]	0.744[m]	0.508	0.424	0.07[u]
Density, lb/gal @ 60°F	6.0-6.5[b]	6.7-7.4[d]	6.63[b]	6.61[b]	6.19[m]	4.22	1.07[r]	–
Boiling Temperature, °F	80-437[b]	370-650[d]	149[c]	172[c]	131[c]	-44	-259	-4,230[u]
Reid Vapor Pressure, psi	8-15[k]	0.2	4.6[o]	2.3[o]	7.8[e]	208	2,400	–
Octane No.[1]								
Research Octane No.	90-100[u]	–	107	108	116[t]	112	–	130+
Motor Octane No.	81-90[s]	–	92	92	101[t]	97	–	–
(R+M)/2	86-94[s]	NA	100	100	108[t]	104	120+	–
Cetane No.[1]	5-20	40-55	–	–	–	–	–	–
Water Solubility, @ 70°F								
Fuel in Water, Volume %	Negligible	Negligible	100[c]	100[b]	4.3[e]	–	–	–
Water in Fuel, Volume %	Negligible	Negligible	100[c]	100[b]	1.4[e]	–	–	–
Freezing Point, °F	-40[g]	-40-30[4]	-143.5	-173.2	-164[c]	-305.8	-296	-435[v]
Viscosity								
Centipoise @ 68°F	0.37-0.44[3,p]	2.6-4.1	0.59[i]	1.19[i]	0.35[i]	–	–	–
Flash Point, Closed Cup, °F	-45[b]	165[d]	52[o]	55[o]	-14[e]	-100 to -150	-300	–

Auto-ignition Temperature, °F	495[b]	600 (approx.)	867[b]	793[b]	815[e]	850-950	1,004	1,050-1,080[u]
Flammability Limits, Volume %								
Lower	1.4[b]	1	7.3[o]	4.3[o]	1.6[e, k]	2.2	5.3	4.1[u]
Higher	7.6[b]	6	36[o]	19[o]	8.4[e, k]	9.5	15	74[u]
Latent Heat of Vaporization								
Btu/gal @ 60°F	900 (approx.)[b]	700 (Approx.)	3,340[b]	2,378[b]	863[5]	775	–	–
Btu/lb @ 60°F	150 (approx.)[b]	100 (approx.)	506[b]	396[b]	138[5]	193.1	219	192.1[v]
Btu/lb Air for Stoichiometric Mixture @ 60°F	10 (approx.)[b]	8 (approx.)	78.4[b]	44[b]	11.8	–	–	–
Heating Value[2]								
Higher (liquid fuel-liquid water) Btu/lb	18,800-20,400	19,200-20,000	9,750[2]	12,800[q]	18,290[h]	21,600	23,600	61,002[v]
Lower (liquid fuel-water vapor) Btu/lb	18,000-19,000	18,000-19,000	8,570[b]	11,500[q]	15,100[h]	19,800	21,300	51,532[v]
Higher (liquid fuel-liquid water) Btu/gal	124,800	138,700	64,250	84,100	–	91,300	–	–
Lower (liquid fuel-water vapor) Btu/ga @ 60°F	115,000	128,400	56,800[3]	76,000[3]	93,500[4]	84,500	19,800[6]	–
Heating Value, Stoichiometric Mixture Mixture in Vapor State, Btu/Cubic Foot @ 68°F	95.2[b]	96.9[5, q]	92.5[b]	92.9[b]	–	–	–	–
Fuel in Liquid State, Btu/lb or Air	1,290[b]	–	1,330[b]	1,280[b]	–	–	–	–
Specific Heat, Btu/lb °F	0.48[g]	0.43	0.6[i]	0.57[i]	0.5[i]	–	–	–
Stoichiometric Air/Fuel, Weight	14.7[3]	14.7	6.45[i]	9[i]	11.7[i]	15.7	17.2	34.3[u]
Volume % Fuel in Vaporized Stoichiometric Mixture	2[b]	–	12.3[b]	6.5[b]	2.7[i]	–	–	–

Notes: Table 1

(1)   Octane values are for pure components. Laboratory engine research and motor octane rating procedures are not suitable for use with neat oxygenates. Octane values obtained by these methods are not useful in determining knock-limited compression ratios for vehicles operating on neat oxygenates and do not represent octane performance of oxygenates when blended with hydrocarbons. Similar problems exist for cetane rating procedures.

(2)   The higher heating value is cited for completeness only. Since no vehicles in use, or currently being developed for future use, have powerplants capable of condensing the moisture of combustion, the lower heating value should be used for practical comparisons between fuels.

(3)   Calculated.

(4)   Pour Point, ASTM D 97 from Reference (c).

(5)   Based on cetane.

(6)   For compressed gas at 2,400 psi.

Source: EIA

(a)   The basis of this table and associated references was taken from American Petroleum Institute, *Alcohols and Ethers*, Publication No. 4261, 2nd. ed. (Washington, D.C., July 1988), Table B-1.

(b)   "Alcohols: A Technical Assessment of Their Application as Motor Fuels," API Publication No. 4261, July 1976.

(c)   *Handbook of Chemistry and Physics*, 62nd ed., 1981, The Chemical Rubber Company Press, Inc.

(d)   "Diesel Fuel Oils, 1987," Petroleum Product Surveys, National Institute for Petroleum and Energy Research, October 1987.

(e)   ARCO Chemical Company, 1987.

(f)   MTBE, Evaluation as a High Octane Blending Component for Unleaded Gasoline," Johnson, R.T., Taniguchi, B.Y., Symposium on Octane in the 1980s, American Chemical Society, Miami Beach Meeting, September 10-15, 1979.

(g)   "Status of Alcohol Fuels Utilization Technology for Highway Transportation: A 1981 Perspective," Vol. 1, Spark-Ignition Engine, May 1982, DOE/CE-56051-7.

(h)   American Petroleum Institute Research Project 44, NBS C-461.

(i)   Lang's Handbook of Chemistry, 13th ed., McGraw-Hill Book Company, New York, 1985.

(j)   "Data Compilation Tables of Properties of Pure Compounds," Design Institute for Physical Property Data, American Institute of Chemical Engineers, New York, 1984.

(k)   Petroleum Product Surveys, Motor Gasoline, Summer 1986, Winter 1986/1987, National Institute for Petroleum and Energy Research.

(l)   Based on isoctane.

(m)   API Monograph Series, Publication 723, "Teri-Butyl Methyl Ether," 1984.

(n)   BP America, Sohio Oil Broadway Laboratory.

(o)   API Technical Data Book - Petroleum Refining, Vol. 1, Chapter 1, Revised Chapter 1 to First, Second, Third, and Fourth Editions, 1988.

(p)   "Automotive Gasolines," SAE Recommended Practice, J312 May 1986, 1988 SAE

Handbook, Vol. 3.

(q)   "Internal Combustion Engines and Air Polution," Obert, E.F., 3rd Ed., Intext Educational Publishers, 1973.

(r)   Value at 80 degrees Fahrenheit with respect to the water at 60 degrees Fahrenheit (Mueller & Associates).

(s)   National Institute for Petroleum and Energy Research, Petroleum Product Surveys, *Motor Gasolines*, Summer 1992, NIPER-178 PPS 93/1 (Bartlesville, Okla., January 1993) Table 1.

(t)   P. Dom, A.M. Mourao, and S. Herbstman, "The Properties and Performance of Modern Automotive Fuels," Society of Automotive Engineers (SAE), Publication No. 861178 (Warrendale, Penn., 1986), p. 53.

(u)   C. Borusbay and T. Nejat Veziroglu, "Hydrogen as a Fuel for Spark Ignition Engines," *Alternative Energy Sources VIII, Volume 2, Research and Development* (New York: Hemisphere Publishing Corporation, 1989), pp. 559-560.

(v)   *Technical Data Book*, Prepared by Gulf Research and Development Company, Pittsburg, Penn., 1962.

Table 2   Comparison of Oxygenate Volume and Weight Percent Oxygen

Oxygen Content (wt%)	MTBE Content (vol%)	TAME Content (vol%)	ETBE Content (vol%)	TAEE Content (vol%)	Ethanol Content (vol%)
2.0	11.0	12.4	12.7	13.0	5.7
2.7	15.0	16.7	17.2	17.9	7.7
3.5	19.2*	22.9*	23.6*	23.5*	10.0

* Oxygenation at these volumes cannot be blended at levels above 2.7 wt% oxygen under current EPA regulations.

Source: Catalytic Distillation Technologies

Table 3 Total Fuel-Cycle Emissions from Gasoline and from Replacement Fuels

Emissions	From Gasoline	From Replacement Fuels, Relative to Gasoline					
		MTBE[1]	Ethanol[2]	LPG	CNG	M85	Electric[3]
Greenhouse Gases[4]							
Water Vapor	Yes	More	More	More	More	More	Less
Carbon Dioxide (CO$_2$)[5]	Yes	Less	Less	Less	Less	Less	a
Carbon Monoxide (CO)	Yes	Less	Equal/Less	Less	Less	Equal	Less
Nitrogen Oxides (NO$_x$)[6]	Yes	Equal	More/Equal[7]	Equal	Equal	Equal	a
Nitrous Oxide (N$_2$O)	Yes	a	a	a	a	a	a
Volatile Organic Compounds[8]							
Methane[9]	Yes	Equal	Equal	Moer	More	Equal	Less
Ethane[10]	Yes	Equal	Equal	Equal	Equal	Equal	Less
Total Ozone Precursors[11]	Yes	Less	More/Less[12]	Less	Less	Less	Less
Nonmethane Organic Compounds							
Methanol	No	More	No	No	No	None	No
Ethanol	No	No	More	No	No	No	No
Formaldehyde	Yes	More	More	Equal	Equal	More	Less
Acetaldehyde	YesLess	More	Less	Equal	Equal	Less	
Sulfur Oxide[13]	Yes	Less	Less	No	No	Less	More
Particulate Matter[14]	Yes	Less	Less	Less	No	No	More

Notes: Table 3

a   Results are uncertain because emissions vary widely, depending on the engine's compression, temperature, and fuel/oxygen mix.

1   Consumed with the gasoline in which it was blended.

2   Includes ethanol in gasohol and ethanol as E85.

3   Life-cycle emissions from electric vehicles depend on the utility feedstock; these projections assume that the feedstock is mostly coal, as more than half of electricity today is generated from coal.

4   Greenhouse gas emission impacts are highly dependent on the feedstock used for fuel production.

5   CO_2 emissions vary widely. In some cases, emissions could be either higher or lower than gasoline, depending on the feedstock and method of production.

6   Nitrogen oxides are not a direct greenhouse gas but rather contribute through tropospheric ozone formation.

7   More for splash-blended gasohol with higher Reid vapor pressure; equal for gasohol with controlled Reid vapor pressure (not splash-blended) and for E85/E100.

8   Methane and ethane are just two of hundreds of volatile organic compounds. Others, such as hydrocarbons from unburned fuel and partial combustion, are not reported here.

9   Does not participate in atmospheric photochemical reactions.

10   Does not participate in atmospheric photochemical reactions.

11   Ozone precursors include NO_x and nonmethane organic compounds.

12   More if splash-blended gasohol but less if specially reformulated gasohol or E85/E100.

13   This catagory includes sulfones. While most greenhouse gases contribute to global warming, the sulfates cool the atmosphere. Among the transportation fuels, sulfur oxides are emitted mostly from diesel fuel.

14   Most particulate matter emissions in the transportation sector come from diesel fuel.

Notes: MTBE, methyl tertiary-butyl ether; LPG, liquified petroleum gas (propane); CNG, compressed natural gas.

Source: EIA

A P P E N D I X   V

STATE ALTERNATIVE FUEL LAWS AND INCENTIVES

(AS OF SUMMER 1995)

For the most up-to-date information on changing state laws and incentives impacting the AFV market, check with the state or local energy office. This list does not include participation in either the oxyfuel or reformulated gasoline programs (See Chapter Three). Also, as a reminder, state and local fleets must meet EPACT and/or CAA requirements.

State	AFV Laws	AFV Incentives
Alabama	None	Up to $75,000 in loans are available to finance AFV projects through the Energy division of the Alabama Department of Economic and Community Affairs.
Alaska	3.5 percent wt oxygen mandate beginning 1996/97 for winter oxy-fuel season in Fairbanks and Anchorage carbon monoxide nonattainment areas	$0.08 per gallon tax incentive for gasoline containing 10 percent ethanol
Arizona	A 1990 state law requires government fleet managers to purchase AFVs if their cost is within 10 percent of the total life cycle cost of a conventionally-fueled vehicle. In 1993, $4.9 million in state funds were set aside to retrofit state fleet vehicles and school buses to help comply with another AFV law that required state fleet managers to replace state vehicles with an alternative fuel (not RFG) at the rate of 10 percent per year beginning in 1993. 1992 legislation gives AFV fleet owners lower vehicle registration fees until Dec. 31, 1998.	Starting in 1994, taxpayers could deduct the following from their gross income tax: *25 percent of the purchase price for one or more new AFVs not to exceed $5,000 *the cost of converting one or more vehicles to operate on alternative fuel, not to exceed $3,000 *the purchase price of refueling equipment for private, noncommercial use, not to exceed $5,000 *50 percent of the interest paid or accrued to purchase an AFV (must be taken in thirds over three taxable years) CNG and LPG are also exempt from the 18 cents per gallon fuel tax.
Arkansas		$250,000 is earmarked for rebates on converting vehicles to AFVs. Up to $1,000 is available for CNG and EV conversions and up to $500 for propane and alcohol (85 percent content or higher).
California	To meet federal clean air goals, the state has adopted the toughest vehicle and fuel emission standards in the country—exceeding the federal standard. California's plan established four progressively stringent standards:	AFVs can receive tax credits of 55 percent of a conversion or retrofit to meet CARB low emission requirements. Maximum write-offs are $1,000 for a light-duty vehicle, $3,500 for vehicles over 5,750

California (Continued)

transitional low-emission vehicles (TLEVs), Low Emission Vehicles (LEVs), Ultra Low Emission Vehicles (ULEVs) and Zero Emission Vehicles (ZEVs). Automakers can participate in a credit trading system that rewards the early introduction of clean fuel/vehicle technology. Currently, automakers anticipate being able to use gasoline-powered vehicles to meet LEV standards and are encouraged that they will meet the ULEV standard.

However, it is the last standard, for ZEV vehicles, that automakers oppose. Even with the credit system in place beginning in the late 1990s, California Air Resources Board regulations demand that up to 10 percent of all vehicles sold by major automakers must be ZEVs.

At this writing, CARB had proposed an E-ZEV (equivalent zero-emission vehicle) standard that would allow electric hybrids (EVs with a small ICE engine) to be considered under the mandate. Credits would also be extended on a one-to-one basis to vehicles with emissions comparable to power plant emissions generated through increased EV demand.

pounds with a total tax credit of $750,000 available to an individual or company through January 1, 1996.

E85 and M85 only pay half of the state's 18 cent per gallon fuel excise tax.

Various municipal areas offer incentives for M85 and E85 powered vehicles. In Sacramento for example, flexible fuel vehicle owners can use up to $500 worth of free M85 fuel if the driver operates 75 percent or more in that district.

Colorado

The state government is required to increase its purchase of AFVs, as a percentage of new vehicle purchases, according to the following fiscal year (FY) schedule:

> FY 1992: 10 percent
> FY 1993: 20 percent
> FY 1994: 30 percent
> FY 1995: 40 percent

Colorado's Air Quality Control Commission adopted Regulation No. 14, establishing emissions performance criteria that after-market conversion systems must meet to be certified for sale in the state. The regulation also provides for the revocation of certification, if warranted, based on the exhaust emissions results from testing in-use vehicles.

On July 1, 1992, Colorado established an alternative fuels financial incentive program to promote the conversion of motor vehicles to run on alternative fuels and to promote the purchase of such vehicles in the state. Under this program, the owner of any new or converted motor vehicle is eligible to apply for financial incentives when the vehicle operates on CNG, propane, electricity, or any other fuel that will achieve comparable emission reduction levels. Financial incentives will be graduated to provide greater rebates for those vehicles that must reduce emissions of CO and "brown-cloud pollutants." The AQCC will

Colorado (Continued)	AQCC Regulation No. 17 implements the state's Clean Fuels Fleet Program for the Denver Metropolitan Area, beginning in 1998. The City and County of Denver required fleets of 30 or more vehicles that are registered in the city to convert 10 percent of their vehicles from gasoline to clean-burning fuels no later than December 31, 1992. Legislation adopted in 1992 established a certification program for mechanics converting motor vehicles to alternative fuels and working on AFV fleets. In 1993, under the "Green Fleets" plan, fuel expenditures must be cut by 1 percent annually, and city fleets must reduce their CO emissions by 1.5 percent, which can be achieved through the purchase of AFVs. Owners or operators of LPG or CNG vehicles pay an annual fuel tax and receive an identifying windshield decal from the vendors and/or distributors of such fuels. Annual fees for the decal range from $70 to $120, depending on vehicle size. Nonprofit transit agencies are exempt from the fuel tax.	promulgate rules to establish the amount of available incentives. In taxable years commencing July 1, 1992, the state has established a tax credit of 5 percent for the purchase of business vehicles using clean-burning alternative fuels or a vehicle converted within 120 days of the date of delivery to use alternative fuels. Clean-burning alternative fuels are defined as "natural gas, LPG, E85, M85, electricity, or any other alternative fuel approved by the AQCC." The credit shall not exceed 50 percent of the actual cost of such conversions and is limited to 50 cars or trucks for each taxable year.
Connecticut	Connecticut is part of a Northeast consortium, the Ozone Transport Commission, which has pledged to adopt California-style exhaust emission standards. The state, however, does not plan to mandate EVs.	There is a $0.01 per gallon credit for gasoline blended with 10 percent ethanol or methanol.
Delaware	AFVs are exempt from the state's 23 cent per gallon fuel excise tax.	
District of Columbia	On December 27, 1990, D.C. enacted the Alternative Fuels Technology Act of 1990 to integrate alternative fuel technology, including participation by the government fleet, commercial transportation fleets, and the Washington Metropolitan Area Transit Authority. On January 4, 1993, the Act was amended to extend the provisions for 2 years beyond the Act's requirements. But due to the city's	

District of Columbia (Continued)

ongoing budget constraints, it is uncertain whether this regulation will be again renewed.

The owner/operator of a fleet must use, through conversion of existing vehicles or procurement of retrofitted or newly manufactured vehicles, a proportional number of motor vehicles powered by clean alternative fuels equal to no less than the following percentages of the entire operational commercial fleet, both public and private:

10 percent by January 1, 1996
15 percent by January 1, 1997
20 percent by January 1, 1998
25 percent by January 1, 1999
30 percent by January 1, 2000
35 percent by January 1, 2001
40 percent by January 1, 2002

Beginning in 1998, no commercial vehicle shall be operated within the Central Employment Area (as defined in the D.C. Zoning Regulations) between sunrise and sunset during the period between May 1 and September 15, unless that vehicle is powered by a clean alternative fuel. On October 1 of each year, each owner and operator of a commercial fleet is required to submit plans to the mayor containing specific goals and timetables for implementation of a clean fuels program.

Florida

By 2000, all state vehicles must operate on the "most efficient, least polluting" alternative fuels.

CNG suppliers are exempt from the $0.123 state fuel tax. The state's energy office has provided $1 million in oil overcharge funds to help state fleet managers meet AFV fleet requirements set under EPACT.

Georgia

The state's Office of Energy Resources has awarded $500,000 in grants for fleet conversions by public entities.

Hawaii		Propane is taxed at two-thirds the rate of diesel fuel tax ($0.16 per gallon). Gasoline containing 10 percent alcohol made from biomass is exempt from the state's 4 percent sales tax on gasoline.
Idaho	State vehicles must use E10 whenever practical.	There is a $0.021 per gallon fuel excise tax exemption on "biofuels" ethanol mixed with gasoline or biodiesel.
Illinois	State vehicles that run on conventional fuels must use gasoline that contains ethanol.	The state offers a 2 percent sales tax exemption for E10 blends.
Indiana		
Iowa	State vehicles that run on conventional fuels must use gasoline that contains ethanol. After 1994, a minimum 10 percent of the state's fleet must run on an alternative fuel.	Iowa offers a $0.01 per gallon excise tax exemption for E10.
Kansas		Ethanol producers receive a direct payment of $0.20 for each gallon of ethanol they produce in the state. The state also offers a $0.03 per gallon gasoline equivalent tax on CNG and propane fuels
Kentucky		
Louisiana	State government fleets are required to phase in purchases of AFVS in the following manner: 1994: 30 percent 1995: 30 percent 1996: 50 percent 1997: 50 percent 1998: up to 80 percent	There is a 20 percent income tax credit for conversion of vehicles and for refueling sites. There is also a reduced rate for CNG and propane fuels.
Maine	As a member of the Ozone Transport Commission, the state intends to adopt California exhaust emission rules, although at present it does not plan to phase in EVs.	
Maryland	Under a 1993 executive order, 20 percent to 25 percent of new state fleet purchases are required to be powered by alternative fuels by 1994. The order	A 1993 law exempts from the sales and use tax the sale of machinery and equipment that convert existing gasoline or diesel fuel-powered

Maryland (Continued)	also establishes an Alternative Fuels Working Group to make recommendations to the governor regarding the purchase and future plansfor AFVs.	motor vehicles to run on clean-burning fuels or intended for use where the fuels are stored, dispensed or delivered. The new law expires June 30, 1999. A 1993 law alters the rate of the motor fuel tax for alternative fuels, as defined under EPACT, from $.2425/gal to $.235/GGE. Another 1993 law exempts from property taxation the machinery or equipment used to dispense clean fuels (that meet the standards of the CAAA) in to motor vehicles. It also provides for the phasing in of the applicability of the property tax to refueling equipment or machinery, beginning in taxable year 1998. The Act applies to all taxable years beginning after June 30, 1994.
Massachusetts	The state has formally adopted the California exhaust emission plan, including a 1998 EV mandate.	The state excise tax on CNG and LPG for on-highway use is about $0.10 per gallon, compared to the $0.21 per gallon excise tax on gasoline.
Michigan		
Minnesota	State vehicles must be fueled with E10 whenever practicable. Chapter 632, Minnesota Laws 1994, redefines ethanol to extend to "residue and waste from the production, processing and marketing of agricultural products, forest products and other renewable resources." Another law provides that the purchase of capital equipment by a contractor for installation in a new ethanol manufacturing facility is exempt from the sales and use tax. Public fleets in the state that operate in a community of 100,000 or moremost convert or purchase new AFVs as follows: FY 1995: 10 percent FY 1996: 15 percent FY 1997: 25 percent FY 1998: 50 percent FY 1999: 75 percent FY 2000: 90 percent	A 1993 law provides $350,000 for the education, outreach, and promotion of ethanol in the state. The state currently provides a 20 cent per gallon incentive for producers of fuel ethanol in the state and a $0.02 per gallon excise tax exemption for 10 vol percent ethanol blended fuels sold in the state. A 1994 law readjusts the producers' incentive from 20 cents per gallon to 25 cents per gallon, starting July 1, 1995, but not to exceed $3.75 million per year per producer. The incentive program's sunset date is extended from 2000 to June 30, 2010. Also included is a phaseout rate of $0.05 per gallon of the $0.02 cent per gallon tax credit, starting October 1, 1994, and

Minnesota (Continued)	Beginning October 1, 1997 all gasoline sold in the state must contain 2.7 percent wt oxygen year-round.	decreasing each year on that date in 1995, 1996, and 1997.
Mississippi		
Missouri	All state vehicles must run on E10. By July 1, 2000, 30 percent of all state vehicles must be AFVs.	The state offers a $0.20 per gallon for instate fuel ethanol producers and a $0.02 per gallon excise tax exemption for E10.
Montana		The state offers a $0.30 producer payment for in-state fuel ethanol production.
Nebraska	State vehicles must run on E10 "whenever practical."	The state's Ethanol Board provides promotion and business development of fuel ethanol industry in the state. There also is a $0.20 per gallon direct producer incentive and a $0.50 per gallon producer payment for ETBE made in the state.
Nevada	AFV requirements are expected for Washoe and Clark counties.	Excise taxes on CNG and LPG motor fuel are $0.18 per gallon compared to the state's $0.22 per gallon tax for gasoline.
New Hampshire	The state is considering California exhaust emission rules.	
New Jersey	The state is considering California exhaust emission rules.	Excise taxes on LPG and CNG motor fuel are $5.25 per gasoline equivalent compared to the $0.105 per gallon tax on gasoline.
New Mexico	State law requires that 30 percent or more of new state and post-secondary institution fleet vehicles operate on alternative fuels effective July 1993, 60 percent in 1994 and 100 percent in 1995 and beyond. It establishes a revolving loan fund to assist financially in the conversion process; it also establishes a state alternative fuel transportation manager. There have been no funds appropriated to date.	The state created a $5 million loan fund for vehicle conversions. A special fuel tax provides for private CNG- and LNG-powered vehicles to pay an annual fee in lieu of the state motor fuel excise tax on vehicles up to 26,000 lb. gross vehicle weight (GVW). To encourage alternative fuels, the state has set a new tax structure for alternative fuels as follows: $0.03/gal from January 1, 1996-December 31, 1997; $0.06 gal from January 1, 1998 to December 31, 1999; $0.09 gal from January 1, 2000 to December 21, 2001; and $0.12/gal after January 1, 2002.

New York	A centrally fueled fleet program, established under law, will require a certain percentage of new clean fuel vehicles to be purchased starting in 1998 and thereafter. A statute adopts California motor vehicle emissions standards, including EVs. New York City Council established a program, under local law 6 (1991), requiring the purchase and/or conversion of AFVs for city government use. The program begins at a 30 percent AFV requirement scaling up to 80 percent of the light-duty nonemergency fleet. Fifteen percent of transit buses are to be alternatively fueled. The city met its goal in 1993 and has over 600 CNG, methanol, and electric cars, trucks and buses. New York established a six-year, $40 million comprehensive AFVFDP to learn the practical requirements of operating AFVs, determine their costs and assist potential users in adapting to the use of these vehicles.	The state provides a partial tax exemption for the sale of new AFVs and for vehicles that are converted to alternative fuel use.
North Carolina		The state offers a personal or corporate income tax credit for new ethanol plant production in the state.
North Dakota	State vehicles must run on E10 " whenever practical."	There is a $0.40 producer payment or ethanol made in the state. The state will fund 10 percent of the cost of conversion up to $200 per vehicle under 10,000 lb, and up to $500 per vehicle for heavy-duty vehicles.
Ohio	State vehicles must run on E10.	There is a $0.01 tax exemption for ethanol blended gasoline.
Oklahoma	The state offers an income tax credit for the fleet vehicle purchase or conversion to alternative fuels, which include CNG, LPG, LNG, methanol, and ethanol. There also is a state-operated fund that reimburses state, county, and municipal governments and school districts that voluntarily convert vehicles.	Instead of a motor fuel excise tax like that used for gasoline or diesel, AFVs are assessed a $100 per vehicle annual fee.

Oregon	State vehicles must run on AFVs "unless not economically available."	An income tax credit has been established for conversion or refueling equipment.
Pennsylvania	The state's energy office offers $1 million in grants for public and private/public partnerships tied to AFV development.	20 percent of a fleet's conversion/retrofit cost is provided by the state.
Rhode Island	The state is expected to adopt California exhaust emission requirements except for EVs.	
South Carolina		
South Dakota	State vehicles must run on E10 "whenever practical."	There is a $0.20 per gallon producer payment for instate ethanol production and a $0.02 excise tax exemption for E10.
Tennessee		
Texas	Local state, federal, private, and school district fleets of 15 or more vehicles, metropolitan rapid transit authorities and city transportation departments must use clean-burning fuels, including RFG if it meets low emission vehicle standards beginning September 1, 1998. Another option available to fleets requires certain percentages of the fleets to meet or exceed the LEV standards, according to the following schedule: September 1, 1998 30 percent September 1, 2000 50 percent September 1, 2002 90 percent Fleet conversion goals may be achieved either by the after-market conversion of existing vehicles or by the purchase of new AFVs. These vehicles may be dedicated, dual-fueled or bi-fueled, capable of operating on conventional or alternative fuels. In 1996, if a planned review shows the program is effective, local government fleets of 15 or more vehicles and private fleets of 25 or more in the state's nonattainment areas will also be required to use alternative fuels according to the following schedule: September 1, 1998 30 percent	The Alternative Fuels Council coordinates alternative fuels programs by state agencies and oversees the Alternative Fuels Conversion Fund, which in part makes loans or grants and finances alternative fuel activities. The Texas Public Finance Authority will be able to sell bonds of up to $50 million for school districts, local mass transit authorities, and state agencies to convert vehicles to alternative fuels, purchase new AFVs and install refueling facilities. Natural gas and propane sold as motor vehicle fuels are exempt from sales tax. Off-road, rental agency, emergency, law enforcement, national security and car dealership vehicles are exempt from the program. The TAFF program also includes a provision for the calculation of mobile emission reduction credits (MERCs) for fleets that meet or exceed the emission and purchase requirements.

Texas (Continued)	September 1, 2000 50 percent September 1, 2001 90 percent Emergency and law enforcement vehicles are exempt from the program's requirements. Waivers from the requirements of Texas' alternative fuel program are available to each affected fleet as specified in state law.	
Utah		The state offers a 20 percent tax credit, (up to $500 maximum) for each dedicated vehicle registered in Utah and a $400 maximum for the cost of conversion equipment if the vehicle is fueled by CNG, propane, or electricity, or is fueled by a fuel that the state's Air Quality Board deems able to meet Clean Air Act goals.
Vermont	Considering California-style exhaust emission standards, including EVs.	
Virginia		CNG, LPG, LNG, hydrogen, or electricity will only be charged a 1.5 percent sales tax as opposed to the 3 percent tax now on gasoline effective January 1, 1996. A state income tax credit of 10 percent is allowed for AFVs and some refueling properties. There is also a job-creation tax credit worth $700 per full-time employee for businesses associated with AFVs for three years. Qualifying businesses include vehicle components and converter manufacturers.
Washington	30 percent of new state vehicles purchased after July 1, 1992 must be LEVs with the purchase requirement increasing each year by 5 percent to reach 70 percent in 2000.	The state offers air pollution control fund matching grants to offset the purchase and operating costs local government shoulder when voluntarily switching to AFVs. In addition, a $750,000 fund has been established for public CNG refueling stations. CNG and LPG vehicles may an annual $85 fee in lieu of fuel excise taxes.

West Virginia	In 1996, 30 percent of state fleet purchases must be AFVs; in 1997, it reaches 50 percent.	
Wisconsin	The governor established an Alternative Fuels Task Force in September1990. The mission of the task force is to evaluate CNG, ethanol, propane, methanol, reformulated gasoline, LNG, electricity, and hydrogen as alternative fuels. The Task Force administers the Local Government Alternative Fuels Cost-Sharing Program, which helped to fund an alternative fuel research laboratory at the University of Wisconsin-Milwaukee, and is developing an Alternative Fuels Economic Development Strategy. Also included is a phased-in program involving propane, CNG, ethanol and biodiesel. State employees must use E10 or alternative fuels for all vehicles leased or owned by the state. The state says it plans to purchase 2,000 AFVs for state use by the year 2000, exceeding EPACT requirements.	The state has initiated public/private partnerships to stimulate ethanol, CNG, propane, and biodiesel fuels infrastructure and use.
Wyoming	State employees must use E10 or alternative fuels for all vehicles leased or owned by the state whenever practical.	E10 has a $0.40 per gallon producer payment for ethanol made in the state.

(Source: DOE/US Alternative Fuels Hotline/Author)

G L O S S A R Y

ACETALDEHYDE — An aldehyde formed as a product of the incomplete combustion in engines using gasoline, methanol, ethanol, propane, or natural gas fuels. Generally, the presence of ethanol or ethyl ethers will lead to acetaldehyde as the primary aldehyde in the exhaust.

AFTER-MARKET — A broad term that applies to any change after the original purchase, such as adding equipment not a part of the original purchase. As applied to alternative fuel vehicles, it refers to conversion devices or kits for conventional fuel vehicles.

AIRSHED — An area or region defined by settlement patterns or geology resulting in discrete atmospheric conditions.

ALCOHOL — ($CH_3(CH_2)n$ -OH) — The family name of a group of organic chemical compounds made of carbon, hydrogen, and oxygen. The series of molecules varies in chain length and is composed of a hydrocarbon, plus a hydroxyl group. Examples of alcohols include methanol, ethanol, and tertiary butyl alcohol.

ALDEHYDES — Aldehydes are toxic chemicals that are products of incomplete combustion in engines using any one of a number of fuels such as traditional gasoline, reformulated gasoline, oxygenated fuel, methanol, ethanol, propane, or natural gas. Alcohol fuels, however, tend to produce higher aldehyde emissions than other fuels due to their chemical similarities. Generally, a fuel that contains methanol or a methanol derivative, such as the oxygenates methyl tertiary butyl ether or tertiary amyl methyl ether, will lead to formaldehyde as the primary aldehyde in the exhaust. Conversely, ethanol or ethyl ethers, like ethyl tertiary butyl ether or tertiary amyl ethyl ether, will lead to acetaldehyde as the primary aldehyde in the exhaust. Both formaldehyde and acetaldehyde are toxic and possibly carcinogenic but at the present the EPA considers them less dangerous than other mobile source emissions like carbon monoxide or nitrogen oxides.

ALKYLATION — A refining process that chemically combines isobutane with olefin hydrocarbons, such as propylene or butylene, through the control of temperature and pressure in the presence of an acid catalyst, usually sulfuric acid or hydrofluoric acid. The product, alkylate, an isoparaffin, has high octane value and can be blended with motor or aviation fuel to boost octane and performance.

ALTERNATIVE FUELS — As defined by the National Energy Policy Act the fuels are methanol; denatured ethanol and other alcohols, separately or in mixtures of 85 percent by volume or more, or other percentage not less than 70 percent as determined by U.S. Department of Energy rule, with gasoline or other fuels; compressed natural gas (CNG);

liquefied natural gas (LNG); liquefied petroleum gas (LPG), propane and hydrogen; coal-derived liquid fuels; fuels other than alcohols derived from biological materials; electricity, or any other fuel determined to be substantially not petroleum and yielding substantial energy security environmental benefits. In practice, an alternative fuel is one that could be used as a replacement motor fuel for traditional gasoline and diesel.

ALTERNATIVE FUEL VEHICLE (AFV) — Motor vehicles that run on fuels other than traditional petroleum-based fuels. As defined by the National Energy Policy Act, this excludes reformulated gasoline as an alternative fuel.

ANODE — A positive electrode (such as in a battery).

AROMATICS — Aromatics, carbon compounds with the carbon strung together in rings, are a common petrochemical building block for gasoline since they have a high energy content. The basic ring consists of six carbon atoms and is shaped like a hexagon. Some heavier aromatics with two or more hexagonal rings with common sides, namely, aromatics, are also present in gasoline, and some are formed during combustion. The U.S. government has sought to restrict their use since some aromatics used in gasoline, like benzene, are carcinogenic. Others, like xylene and some of the more complex aromatics, are active ground-level ozone formers.

BENZENE (C_6H_6) — A carcinogenic aromatic hydrocarbon present in small proportions in some crude oils and as a component of high-octane gasoline. It can be made commercially from petroleum by the catalytic reforming of napthenes in petroleum naphtha. Also made from coal in the manufacture of coke. In addition to its use in gasoline, it can be used as a solvent, in manufacturing detergents, synthetic fibers, and as a feedstock for other petrochemicals.

BIFUEL VEHICLE — A vehicle with two separate fuel systems designed to run on either an alternative fuel or conventional gasoline, using only one fuel at a time. These systems are advantageous for drivers who do not always have access to an alternative fuel refueling station. Bifuel systems are usually used in light-duty vehicles.

BIODIESEL — A biodegradable transportation fuel for use in diesel engines produced through the transesterfication of organically-derived oils such as soybeans, rapeseed, or sunflowers, with an end product known as methyl ester, in which case the animal tallow, end product is methyl tallowate. Biodiesel can also be made by transesterfication of hydrocarbons produced by the Fisher-Tropsch process from agricultural byproducts like rice hulls. It may be used either as a replacement for or as a component of diesel fuel.

BIOFUELS — Wood, waste, or alcohol fuels made from an agricultural or renewable resource, such as corn, wheat, or grass.

BRITISH THERMAL UNIT (Btu) — A standard unit for measuring heat energy. One Btu represents the amount of heat required to raise one pound of water one degree Fahrenheit at sea level.

BUTANE (C_4H_{10}) — A typically gaseous straight-chain or branch-chain hydrocarbon extracted from natural gas or refinery gas streams. In a production process, it would include isobutane and normal butane.

BUTYLENE (C_4H_8) — An olefinic hydrocarbon recovered from refinery processes. It is often used as a feedstock for oxygenates used in gasoline including methyl tertiary butyl ether (MTBE) and its chemical sister ETBE (ethyl tertiary butyl ether).

CALIFORNIA AIR RESOURCES BOARD (CARB) — A state regulatory agency which regulates air quality in California. Air quality regulations established by this independent agency are often stricter than those set by the federal government.

CARBON DIOXIDE (CO_2) — A colorless, odorless, non-poisonous gas that is a normal part of the ambient air and is converted into energy by plants through photosynthesis. CO_2 is also a product of fossil fuel combustion. Although CO_2 does not directly impair human health and is a key part of the atmosphere, it is also a greenhouse gas that when present in elevated levels can trap the earth's heat and contribute to global warming if left unchecked.

CARBON MONOXIDE (CO) — A colorless, odorless gas slightly lighter than air. It is poisonous if inhaled, in that it combines with blood hemoglobin to prevent oxygen transfer. It is produced by the incomplete combustion of fossil fuels, and emission levels are particularly higher in older automobiles with less efficient emission control devices.

CATALYTIC REFORMING — A refining process using controlled heat and pressure with catalysts to rearrange certain hydrocarbon molecules, thereby converting paraffinic and naphthionic type hydrocarbons, such as low-octane gasoline boiling range fractions, into petrochemical feedstocks and higher octane stocks suitable for blending into finished gasoline.

CERTIFICATION — A process by which a motor vehicle, motor vehicle engine, or motor vehicle pollution control device satisfies the criteria adopted by a state or federal regulatory agency. Certification constitutes a guarantee by the manufacturer that the engine will meet certain standards at a certain mileage level, often 50,000 miles; if not, it must be replaced without charge.

CLEAN FUEL VEHICLE — Incorrectly used interchangeably with alternative fuel vehicle. Generally, refers to vehicles that use low-emission, clean-burning fuels. In California, Public Resources Code No. 25326 defines clean fuels, for purposes of the section only, as fuels designated by the California Air Resources Board for use in low, ultra low, or zero emission vehicles, including but not limited to electricity, ethanol, hydrogen, liquefied petroleum gas, methanol, natural gas, and reformulated gasoline.

CLUNKERS — Also known as gross-polluting or super-emitting vehicles. Vehicles that emit far in excess of the emission standards by which the vehicle was certified when it was new.

COMPRESSED NATURAL GAS (CNG) — Natural gas that has been condensed under high pressure, typically between 2,000 and 3,600 pounds per square inch, held in a container. The gas expands when released for use as a fuel.

CONVERSION — A device or kit by which a conventional fuel vehicle is changed to an alternative fuel vehicle.

CONVERTED VEHICLE — A vehicle originally designed to operate on gasoline, which has been modified or altered to run on an alternative fuel.

CORPORATE AVERAGE FUEL ECONOMY (CAFE) — A sales-weighted average fuel mileage calculation, in terms of miles per gallon, based on city and highway fuel economy measurements performed as part of the federal emissions test procedures. CAFE requirements were instituted by the Energy Policy and Conservation Act of 1975 and modified by the Automobile Fuel Efficiency Act of 1980. For major manufacturers, CAFE levels are currently 27.5 miles per gallon for light-duty automobiles. CAFE standards also apply to some light trucks. The Alternative Motor Fuels Act of 1988 allows for an adjusted calculation of the fuel economy of vehicles that can use alternative fuels, including fuel-flexible and dual-fuel vehicles.

DEDICATED VEHICLE — A vehicle designed to operate solely on one alternative fuel.

DUAL-FUEL — A vehicle capable of operating on two different fuels, in distinct fueling systems, such as compressed natural gas and gasoline.

ELECTRIC VEHICLE (EV) — A vehicle powered by electricity, usually provided by batteries but may also be provided by photovoltaic cells or a fuel cell.

EMISSION CREDIT TRADING — A program administered by the EPA under which low polluters are awarded credits which may be traded on a regulated market and purchased by polluters who are in noncompliance for emissions until compliance can be achieved.

E10 — A mixture of 90 percent by volume gasoline and 10 percent by volume ethanol.

E85 — A mixture of 85 percent by volume ethanol, 15 percent by volume unleaded gasoline used in flexible fuel vehicles.

ETHANOL (CH_3CH_2OH) — Also know as ethyl alcohol or grain alcohol. A liquid produced chemically from ethylene or biologically from the fermentation of various sugars from carbohydrates found in agricultural crops and cellulosic residues from crops or wood. Used in the United States as a gasoline octane enhancer and oxygenate, it increases octane 2.5 to 3 numbers at 10 percent concentration. Ethanol can also be used in higher concentration in vehicles optimized for its use.

ETHYL TERTIARY BUTYL ETHER (ETBE) — An aliphatic ether similar to MTBE. This fuel oxygenate is manufactured by reacting isobutylene with ethanol. Having high octane and low volatility characteristics, ETBE can be added to gasoline up to a level of approximately 17 percent by volume. ETBE is used as an oxygenate in some reformulated gasolines.

FLEXIBLE FUEL VEHICLE (FFV) — A vehicle that can operate on either alcohol fuels or regular unleaded gasoline or any combination of the two from the same tank.

FORMALDEHYDE (HCHO) — An aldehyde formed as a product of the incomplete combustion in engines using gasoline, methanol, ethanol, propane, or natural gas fuels. Generally, the presence of methanol or methyl ethers will lead to formaldehyde as the primary aldehyde in the exhaust.

FUEL CELL — An electrochemical engine with no moving parts, which converts the chemical energy of a fuel, such as hydrogen, and an oxidant, such as oxygen, directly into electricity. The principal components of a fuel cell are catalytically activated electrodes for the fuel and the oxidant and an electrolyte to conduct ions between the two electrodes, thus producing electricity.

FUNGIBLE — A term often used within the oil refining industry to denote products that are suitable for transmission by pipeline. Ethanol, for example, is not considered fungible in this sense, in that it would absorb any water accumulating in pockets in a pipeline.

GASOHOL (E10) — In the United States, a gasoline that contains 10 percent ethanol by volume. This term was used in the late 1970s and early 1980s but has been replaced in some areas of the country by E10, Super Unleaded Plus Ethanol, or Unleaded Plus.

GREENHOUSE EFFECT — A scientific theory used to describe the roles of water vapor, carbon dioxide, and other trace gases in keeping the Earth's surface warmer than it would be otherwise. These radioactively active gases are relatively transparent to incoming shortwave radiation, but are relatively opaque to outgoing longwave radiation. The latter radiation, which would otherwise escape to space, is trapped by these gases within the lower levels of the atmosphere. The subsequent reradiation of some of the energy back to the Earth maintains surface temperatures higher than they would be if the gases were absent.

HYBRID VEHICLE — Usually a vehicle which employs a combustion engine system together with an electric propulsion system. Hybrid technologies expand the usable range of electric vehicles beyond what an all-electric vehicle can achieve with batteries only.

HYDROGEN (H_2) — The lightest of all gases, this element occurs chiefly in combination with oxygen in water. It also exists in acids, bases, alcohols, petroleum, and hydrocarbons, and can be used in motor transportation in a fuel cell or liquefied.

INFRASTRUCTURE — Generally refers to the recharging and refueling network necessary for successful development, production, commercialization, and operation of alternative fuel vehicles, including fuel supply, public and private recharging and refueling facilities, standard specifications for refueling outlets, customer service, education and training, and building code regulations.

Inherently Low Emission Vehicle (ILEV) — A term used by the federal government for any vehicle that is certified to meet the California Air Resources Board Transitional Low Emission Vehicle (TLEV) standards and does not emit any evaporative emissions.

ISOBUTYLENE (C_4C_8) — An olefinic hydrocarbon recovered from refinery processes or petrochemical processes. See also Butylene.

ISOMERIZATION — A refining process that alters the fundamental arrangement of atoms in a molecule without adding and removing anything from the original material. This process is used to convert normal butane into isobutane (C_4), an alkylation process feedstock, and normal pentane and hexane into isopentane (C_5) and isohexane (C_6), high octane gasoline components.

LIQUEFIED NATURAL GAS (LNG) — A natural gas that has been condensed to a liquid, typically by cryogenically cooling the gas to -327.2 degrees Fahrenheit.

LIQUEFIED PETROLEUM GAS (LPG) — A mixture of gaseous hydrocarbons, mainly propane and butane, that change into liquid form under moderate pressure. LPG, or propane, is commonly used as a fuel in rural homes for space and water heating, as a fuel for barbecues and recreational vehicles, and as a transportation fuel. It is normally created as a byproduct of petroleum refining and from natural gas production. California does not consider propane as an alternative to petroleum because of its direct links to oil, though it is listed in federal regulations and laws as an alternative fuel.

LOW EMISSION VEHICLE (LEV) — A vehicle certified by the California Air Resources Board to have emissions from zero to 50,000 miles no higher than 0.075 grams/mile (g/mi) of nonmethane organic gases, 3.4 g/mi of carbon monoxide, and 0.2 g/mi of nitrogen oxides. Emissions from 50,000 to 100,000 miles may be slightly higher.

M85 — A blend of 85 percent methanol and 15 percent unleaded gasoline, used as a motor fuel in flexible fuel vehicles.

M100 — 100 percent methanol, used in dedicated methanol vehicles such as heavy-duty truck engines.

METHANE (CH_4) — The simplest of hydrocarbons and the principal constituent of natural gas. Pure methane has a heating value of 1,1012 Btus per standard cubic foot.

METHANOL (CH_3OH) — Also known as methyl alcohol or wood alcohol. A liquid formed by catalytically combining carbon monoxide with hydrogen in a 1:2 ratio, under high temperature and pressure. Commercially it is typically made by steam reforming natural gas. Also formed in the destructive distillation of wood.

METHYL TERTIARY BUTYL ETHER (MTBE) — An ether manufactured by reacting methanol and isobutylene. The resulting ether has a high octane and low volatility. MTBE is a fuel oxygenate and is permitted in unleaded gasoline up to a level of 15 percent. It is one of the primary ingredients in reformulated gasolines.

NATURAL GAS — A mixture of hydrocarbon compounds and small quantities of various nonhydrocarbons existing in the gaseous phase or in solution with crude oil in natural underground reservoirs.

NATURAL GAS VEHICLE (NGV) — Vehicles that are powered by compressed or liquefied natural gas.

NONATTAINMENT AREA — A region that exceeds the minimum acceptable federal clean air standards as set by the EPA. Such regions are required to change an EPA-approved clean air, State Implementation Plan, in order to meet the standard in an "acceptable" timeframe. Under the Clean Air Act, if a nonattainment area continues to fail the standard, EPA may superimpose its own Federal Implementation Plan, which could include stricter requirements on auto emission standards, imposing of fines, construction bans, or other cutoff in federal aid.

OFF-ROAD — Any nonstationary device powered by an internal combustion engine or motor, used primarily off the highways to propel, move, or draw persons or property, including any device propelled, moved, or drawn exclusively by human power, and used in any of the following applications: marine vessels, construction or farm equipment; locomotives; utility, lawn, and garden equipment; off-road motorcycles; and off-highway vehicles.

ORIGINAL-EQUIPMENT MANUFACTURER (OEM) — A manufacturer of complete vehicles or heavy-duty engines, as contrasted with remanufacturers, converters, retrofitters, and repowering or rebuilding contractors overhauling engines, adapting or converting vehicles or engines obtained from the OEMs, or exchanging or rebuilding engines in existing vehicles.

OXYGENATE — A term used in the petroleum industry to denote octane components containing hydrogen, carbon, and oxygen in their molecular structure. Includes ethers, such as MTBE and ETBE, and alcohols, such as ethanol or methanol. The oxygenate is a prime ingredient in reformulated gasoline. The increased oxygen content given by oxygenates promotes more complete combustion, thereby reducing tailpipe emissions.

OZONE (O_3) — An oxygen molecule with three oxygen atoms, which occurs as a blue, harmful, pungent-smelling gas at room temperature. The ozone layer, a concentration of ozone molecules located six to 30 miles above sea level, is in a state of dynamic equilibrium that absorbs much of the ultraviolet radiation of the sun, shielding life from the harmful effects of radiation. Ozone at ground-level concentrations is a dangerous air pollutant and can be caused through a series of complex reactions with various pollutants, usually found in urban settings. In addition, some pollutants can drift up into the upper atmospheres and damage the ability of ozone to absorb ultraviolet radiation.

OZONE PRECURSOR — A chemical compound, such as nitrogen oxides, methane, nonmethane hydrocarbons and hydroxyl radicals, that in the presence of solar radiation, reacts with other chemical compounds to form ground-level ozone pollution. Tailpipe emissions from cars are a leading ozone precursor source.

PARTICULATE MATTER (PM) — Unburned fuel particles that form smoke or soot and stick to lung tissue when inhaled. A chief component of exhaust emissions from heavy-duty diesel engines.

PROPANE — See Liquefied Petroleum Gas.

RATE-BASING — Refers to practice by utilities of allotting funds invested in utility research and development, demonstration, and commercialization, and other programs from ratepayers, as opposed to allocating these costs to shareholders.

REFORMULATED GASOLINE (RFG) — Gasolines that have had their compositions and/or characteristics altered to reduce vehicular emissions of pollutants. To meet federal RFG requirements, a fuel must include oxygenates, have reduced olefin and aromatic content, and meet performance specifications for ozone-forming emissions and toxic substances.

REID VAPOR PRESSURE (RVP) — A standard measurement of a liquid's vapor pressure in pounds per square inch at 100 degrees Fahrenheit. It is an indication of the propensity of the liquid to evaporate.

RENEWABLE ENERGY — Energy obtained from sources that are essentially inexhaustible. Renewable energy sources include hydroelectric, wood, waste, geothermal, wind, photovoltaic, and solar thermal energy.

REPLACEMENT FUEL — A portion of any motor fuel that is methanol, ethanol, other alcohols, natural gas, liquefied petroleum gas, hydrogen, coal-derived liquid fuels, or electricity, including electricity from solar energy, that is not petroleum-based.

RETROFIT — A broad term that applies to any change after the original purchase, such as adding equipment not a part of the original purchase. As applied to alternative fuel vehicles, it refers to conversion devices or kits for conventional fuel vehicles. Essentially the same as AFTER-MARKET.

TERTIARY AMYL ETHYL ETHER (TAEE) — An ether based on reactive C5 olefins and ethanol.

TERTIARY AMYL METHYL ETHER (TAME) — An oxygenate that can be used in reformulated gasoline. It is an ether based on reactive C5 olefins and methanol.

TRANSITIONAL LOW EMISSION VEHICLE (TLEV) — A vehicle certified by the California Air Resources Board to have emissions from zero to 50,000 miles no higher than 0.125 grams/mile (g/mi) of nonmethane organic gases, 3.4 g/mi of carbon monoxide, and 0.4 g/mi of nitrogen oxides. Emissions from 50,000 to 100,000 miles may be slightly higher.

ULTRA-LOW EMISSION VEHICLE (ULEV) — A vehicle certified by the California Air Resources Board to have emissions from zero to 50,000 miles no higher than 0.04 grams/mile (g/mi) of nonmethane organic gases, 1.7 g/mi of carbon monoxide, and 0.2 g/mi of nitrogen oxides. Emissions from 50,000 to 100,000 miles may be slightly higher.

WARRANTY — A seller's guarantee to purchaser that a product is what it is represented to be and if it is not, that it will be repaired or replaced. Within the context of vehicles, refers to an engine manufacturer's guarantee that the engine will meet certified engine standards at 50,000 miles or the engine will be replaced. Retrofits will generally void an engine warranty.

ZERO EMISSION VEHICLE (ZEV) — Any vehicle that is certified by the California Air Resources Board to have zero tailpipe emissions. The only vehicles that currently qualify at ZEVs are electric vehicles. Hydrogen-powered vehicles could also meet the standard, but none exist on a commercial level at the present time.

I N D E X